Paul Ansel Chadbourne

Lectures on Natural History

Its Relations to Intellect, Taste, Wealth and Religion

Paul Ansel Chadbourne

Lectures on Natural History
Its Relations to Intellect, Taste, Wealth and Religion

ISBN/EAN: 9783337026301

Printed in Europe, USA, Canada, Australia, Japan

Cover: Foto ©berggeist007 / pixelio.de

More available books at **www.hansebooks.com**

LECTURES

ON

NATURAL HISTORY.

———◆◆◆———

CHADBOURNE.

LECTURES

ON

NATURAL HISTORY:

ITS RELATIONS

TO

INTELLECT, TASTE, WEALTH, AND RELIGION.

BY

P. A. CHADBOURNE,

PROFESSOR OF NATURAL HISTORY IN WILLIAMS COLLEGE,
AND
PROFESSOR OF NATURAL HISTORY AND CHEMISTRY
IN BOWDOIN COLLEGE.

NEW YORK:

A. S. BARNES & BURR, 51 & 53 JOHN-STREET.

1860.

RENNIE, SHEA & LINDSAY,
STEREOTYPERS AND ELECTROTYPERS,
81, 83 & 85 CENTRE-STREET,
NEW YORK.

GEO. W. WOOD, PRINTER,
No. 2 Dutch-st., N. Y.

PREFACE.

It is a characteristic of the American people, to test every thing by its money value alone. The brief discussion of the Relations of Natural History in the following Lectures, was entered upon with the hope of doing something to show that this department of study is by no means to be estimated by its direct return of dollars and cents. Simply to impart information, is a small part of the teacher's work. This is not to be neglected; but training the mind, so that it shall move on, a living, expanding power through life, is *education*. As the living tree gathers with its thousands of rootlets nutriment from the earth beneath, while its leaves are drawing in the gases from every breeze that moves them, to build up the fabric—so the mind must be trained to gather food from every field of thought, and change, by its vital power, to an element of strength, the mental accumulations which to many become a burden to the memory alone. Many students, enjoying a high reputation for accuracy, leave college with a knowledge of the text-books indeed, but, we might almost say, unfitted for future acquisitions by those already made. That studies in an educational course should be selected for their educating power, would seem to be evident. But the truth is, *information* is mistaken for *education*. And Natural History has in general been valued simply for the information it furnishes, rather than as an educating power. It is in this light that it is generally matched against the Dead Languages. We wish to put it in the place of no other study, certainly not in the place of the Ancient Languages or Mathematics, without both of which its profitable study is almost hopeless. We simply wish to claim for it a higher rank than has thus far been assigned to it, by showing its varied relations to man.

The study of a single term, or a brief course of lectures, has generally been considered sufficient for the great book of Nature, while two or three years are required on ancient languages before commencing the collegiate course. So that while almost every graduate considers himself competent to teach Latin, Greek, or Mathematics, probably not one in ten would offer himself as qualified to instruct in Natural History.

The Lectures are printed as prepared for

delivery, either as a course or separately, although this makes repetition unavoidable. Nothing, perhaps, would be gained by attempting to avoid this. It is hoped that they may at least prove acceptable to those who have listened to them, and to those who honestly ask the question, "What is the use of the Naturalist's work?"

BOWDOIN COLLEGE, 1860,

CONTENTS.

THE RELATIONS

OF

NATURAL HISTORY.

LECTURE I.

NATURAL HISTORY AS RELATED TO INTELLECT.

ON the banks of the Tigris there is the palace of
a king who has no successor among the living mon-
archs, and his subjects have long since ceased to be
reckoned among the powers of the world. For
more than two thousand years earth and rubbish
have covered its ruined walls, and filled its winding
galleries that once echoed to the tread of busy life.
Its site even became unknown to those who pitched
their tents by its side, or buried their dead in the
mound that inclosed its foundations. But this
burying-place of former grandeur, and of the pass-

ing generations, has not been left undisturbed.
From their resting-places have been brought up the
slabs that in a measure reveal the thought of this
ancient people. The king has engraven his name
on the back of the slabs that form a part of his palace, while upon their fronts his mighty acts are
chiseled in the cuneiform characters of his nation.
Even the clay tile has stamped upon it some name
or story. Why is it that those huge blocks of stone
are sawn asunder, floated down the Tigris or transported on camels' backs, and then borne across the
ocean to take their places in our museums? Do we
expect, like their makers, that these old divinities
will give fruits to the field, and victory in war?
We do not believe they have power to save or to
destroy. Why do scholars bend with wearied eye
and throbbing brain over these old mutilated inscriptions? Do they expect to find in them lessons
of wisdom which they have never read in other languages? or to make, by such labor, discoveries in
art and science, which shall lengthen human life,
alleviate its ills, or add to its comforts? None of
these things are expected. The old deities are to us

mere stone—brittle slabs of mingled clay and gypsum; their mystic cones meaningless, their carving
uncouth, their inscriptions some idle vaunt of vain-
glorious kings, only equaled by the senseless self-
laudations of the "Brother of the Sun." But in
every line upon those old marbles there is the
record of a thought; and whatever its value or
worthlessness, we wish to throw its light on the
great background of human history. It is the
search for thought that dignifies the labor among
the mounds of Nineveh,—that redeems it from the
charge of childish folly, and makes each new discovery a matter of universal interest. It is not the
value of the new stone, nor the value of its inscription in bringing to light new views in morals or
philosophy, nor new facts in science; but there is
there another thought, collected rays of thought in
the figure, the position of the marble, and in its inscription, that can together throw light on the great
historic perspective where the converging pillars are
lost in darkness. It is thus, and for this reason, that
we seek to gather from the mounds of our own
country the relics of a lost people. We gather their

2

rude implements; even the broken pottery is a treas-
ure: and all this to pierce the curtain of mystery
that hangs over their origin and history—to catch a
glimpse, if possible, of some broken shaft in that
long gallery of history, which fell so long before
Columbus lived, that not a single arch has been
borne to us on the bosom of Indian tradition to aid
us in its reconstruction.

This is natural to man. Whatever gives evidence
of thought, he wishes to investigate. The field of
thought is the home of a thinking being, the home
of man; and whatever manifests thought, without
evil associations, is never by him to be regarded as
useless. He never can thus regard it, for the very
law of his intellectual being forbids it. He may
not have so far analyzed his intellectual forces as to
know why he is impelled to this or that investigation. He may not be able to give a satisfactory
answer to the one who demands the use. But, he
knows there is a use, as he knows that food strength-
ens his body, although he may be in happy ignorance of such an organ as a stomach, and have no
notion of the peculiar office of carbon and nitrogen

compounds. He can not tell how the food acts, but he goes on eating, for his appetite demands it. In satisfying its cravings, the good of the body is cared for. It was given to guide men, before science could help them. It led them in the right direction as surely before the days of Hunter and Liebig, as it does now with all the light of modern science. So this intellectual appetite, that has led men to dig among ruins, to wipe the dust from the ancient inscription, to gather as a pearl every monument of human thought, to scan every form of matter as it exists in nature—the crystal and the flower—the animal, from the largest to the animalcule—those now living and those sleeping in their beds of stone —this intellectual appetite has led men in the right direction. It has led them to labor, though unable to defend themselves from sneers, and unable to frame arguments in favor of what they knew must be right.

It is this fact in Natural History—its manifestations of thought—that has enchained so many brilliant intellects in its pursuit, from the days of Aristotle till the present time. This was the charm that bound them to their work, and cheered them in their investigations. The power of this element has never been more fully recognized than in the late work of the great master in Zoology, who sums up each of his thirty-three first chapters as expressions of thoughts of the Creator. He does not, like the Alchemist, claim that he has made the gold which he holds up to our admiring view. He presents the gleaming ore, and says, Here I found it, where it was poured in all its purity by God himself.

We have now laid open broad veins by centuries of patient search; but it was the shining particles of the same true ore, the thought of God, that led on the early searchers, though they found it in grains so small and scattered, while walking upon the edge of the placer, that the multitude could see nothing. We have drawn on to richer fields, and Natural History has assumed such an importance—so many are engaged in its pursuits—it is coming to take such a place in our courses of instruction—that we may well inquire its relations to man as an intellectual, emotional, physical, and religious being; or, in other words, the relation of

Natural History to INTELLECT, TASTE, WEALTH, AND
RELIGION.

Its relations to *wealth* are most generally considered by the common people, and even by those who
are clamorous that it should take the place of other
subjects in our courses of study. Its study is not
demanded by them because they believe it better
fitted than Euclid, or Horace, or Thucydides, as a
discipline to the mind, but they see that this may
be a road to wealth in a country like ours, abound-
ing in mineral riches. That its money value, on
short time, is the ground of their estimation, is apparent from the fact that they are eager for so
much of the study as relates to mines, while it
seems as ridiculous to them as ever that men
should dissect fishes, catch bugs and butterflies, or
worse still, write whole books on turtles' eggs.
From the selection they are sure to make from the
departments of Natural History, we have a key by
which to translate their common question, " What
is the use of it?" It is simply this, how much
ready-money will it bring ? Will it bring in more
money than bank-stock or government five-per-

2ᵃ

cents? "I wish," argues the prudent father, "to
give my son one thousand dollars; it must be safely
invested, so that it will bring in sixty dollars an-
nually. Shall I put it into the vault of the sav-
ings-bank, or into my son's head in the shape of
Natural History?" It is with him a mere matter
of judicious investment. The son's head is balanced
against the stone vault, or a wooden box, as a safe
place for depositing money. If the box is surest to
bring semi-annual dividends, the money goes there,
and the apartments in the son's head are still empty.

The argument from design is so obvious, and has
been so well presented, that a certain relation of
Natural History to religion is acknowledged by
those who have given the least thought to this great
revelation. That it has other and more important
bearings than these special arguments thus far presented, it would not be difficult to show.

But two important departments still remain—*Intellect* and *Taste*—that have not yet been properly
connected with Natural History, so that it should
be seen to have high claims in reference to them
alone. On the first of these, the relation of Natural

History to *Intellect*, I propose to speak at this time.
And we trust it will appear that in this view alone,
it fully justifies the enthusiasm and the labor of naturalists in all ages, and will justify their continued
labor, until every object in nature is searched out,
and the thought in it revealed.

I need not stop here to prove that the intellect is
to man more than money—that money can be only
a means of accomplishing good, while the cultivated
intellect is not only a means, but is itself an end,
a positive good; because, by its exercise, man rises
constantly to greater capacities for enjoyment by the
very act of enjoying. Its revenue is unalloyed with
the anxieties that wealth necessarily brings. By
the intellect men may rise so high, that neither
wealth nor station can add any thing to their influence, and poverty can take nothing from it, nor
lessen the respect in which they are held. We
never think of wealth in connection with Newton,
Cuvier, nor Humboldt. They are in a sphere so
high, that neither riches nor poverty are known or
recognized there. Official station could not lend
them dignity. Nothing but immorality could shake

the intellectual thrones which they occupied.   One
of them studied the heavens; another brought living forms from the dead bones of Montmartre; the
other has scanned the various aspects of nature in
every clime, and still lives a companion of kings,
and an honor to the race.*   They were not wholly
made what they were by the study of nature, but
the giant intellects which God gave them, found
in the study of nature adequate employment and
means of perpetual growth.   They walked with nature, as the scholar walks with the great master, list-
ening as he unfolds his thoughts, and deferentially
propounding questions in every case of doubt.   It
was because in nature there was thought embodied
—the constant unfolding of a plan drawn by infinite
wisdom, and written out on every star and mountain
—in all the tribes of land and water—in the expanding flower and glittering grain of sand—that they
never tired of her communings, never grew wiser
than their teacher, but felt themselves to be children
to the last.

* Deceased since this was written.

The lives of such men—men never to be spoken of but with admiration—would be enough, one might think, to insure the study of nature from neglect. But this general assent which their commendation might imply, is still withheld from a multitude of objects on which naturalists spend their lives. Newton, we may be told, was an astronomer, and that walking among the stars is a very different thing from groping in mud and water for the puny objects of Natural History. It is true, Newton seems to us, now, always surrounded with a halo of stellar light; but when on earth he excited the compassion of his neighbors by his, to them, senseless employment of blowing soap-bubbles. How that act has become dignified in the opinion of men by the results which have flowed from it! It was to Newton then, more than it can be to them now. He saw in the prismatic colors of the trembling bubble, laws of matter wonderful in their possible results—with all the charm of novelty, if we can apply this tame expression *novelty* to that happy emotion which calls pleasure from every fiber of the intellectual being when a new relation, or law of nature, flashes upon the mind. The time will come, when the humblest work of the Natural Historian will, like the soap-bubble of Newton, vindicate itself. We are sure this is so; for in every created object—in the myriad forms thrown up by every wave—in the beetle that fills with drowsy hum the evening air—in the worm that crawls—in the moss and mildew—there is a thought of God. We are sure, that what it was not beneath the dignity of God to create, is not beneath the dignity of man to study—that it can not fail to vindicate fully its claim to our attention. We simply wish to do something to hasten that time.

Natural History is the study of the earth as one mass, and of every object upon its surface and within its crust. We ask you, then, to enter the portals of this great temple, and read the thought of the Builder in every separate stone, and its joining. Nothing is superfluous—nothing is wanting. Every line, seemingly useless in the separate stones, serves to show their true place in the arch or dome. And not a single tint could be lost without marring the grand picture which the pieces all conspire to form.

They are like the colored glass of some grand old cathedral window—forming a picture unseen by those who pass on the outer side of the temple, but to those within, giving gorgeous tints and celestial groups.

We spend days and nights in our libraries, communing with the great of the past ages—and we do well. It gives strength and beauty to the mind to drink in the thoughts of those who towered up as beacon-lights to the world. We make long journeys to see the works of the great masters; but in this temple of nature, which opens her portals to us in every land, we commune with Him who "by wisdom hath founded the earth."

We step first into the lowest vestibule of this temple—the mineral kingdom. And here, as will subsequently appear in the higher departments, we may examine each object independently, or we may direct our attention to the grouping of the whole—the relation of each object to others, or the relation of the whole to higher departments. In this examination it will be impossible to keep entirely clear of related sciences, as Chemistry and Meteorology.

All the natural sciences are so joined that no one of them can be properly considered without some aid from others; or, at least, by so far introducing them as to show the line of junction, as adjacent territory is generally drawn in outline around any portion of the earth that we wish to map with precision.

We have learned by the aid of Chemistry that there are sixty-two kinds of matter. All of these elements occurring in a simple state, and the compounds of the whole number existing as natural products, belong to this one lowest department of Natural History—Mineralogy. It is the same matter indeed as is found in the higher departments, but it is combined and controlled by inferior forces; chemical affinity being the highest force ever manifested in a mineral. We have here hundreds of substances making up the earth's crust, mingled in seeming confusion, and many of them of-protean form. These are to be sought out, and their true nature discovered under their various disguises. Were there no plan nor law in their structure, the task would be hopeless. For where there is no relationship, the study of one object can give no aid in understanding another. Any arrangement not founded upon like nature is only an arbitrary placing, which is no sign of progress in any department. But these have each a definite plan, and each a relationship to some other. And upon them are stamped the characters by which their nature may be known, by those who look with patient study. There is engraven within their very structure a story, an autobiography that unrolls the more the longer we gaze upon it. It is perfect, for the writing is a transcript, by their maker, of the nature He has given them;—not like the daguerreotype, the very shadow, but the very thing itself. It is the nature given by God, manifested in all those sensible signs by which the thing is known.

A celebrated mineralogist was once asked how he knew that a certain body had fallen from the heavens, which he was giving thousands of dollars for, to enrich his collections of meteorites. His answer was, "I see the finger-marks of the Almighty stamped upon every part of it." This might seem a bold expression, or as indicating some wonderful property in those bodies that fall from the heavens. But if

such language could be applied to a meteorite, it is
equally true of every pebble beneath our feet. To
translate these marks, to read this language of the
mineral kingdom, we have in kind the highest conditions for mental activity. Other departments may
give us higher *degrees*. **We have here a multitude**
of forms — each form perfectly defined — sensible
properties varied without limit — all combined form-
ing labels for every species in the mineral kingdom,
as perfect as the works of God ever are, and yet only
to be read by the keenest mental insight, and by
calling into active exercise every sense. The nature
of this language we have already indicated; but we
will examine it more in detail, because it is that in
which the whole book of nature is written. And he
who would in after-life read the inscriptions on her
grand old arches — the poems in her grottoes — must
not despise the alphabet which, meaningless by itself, is the only key to unlock those well-springs **of**
knowledge which the multitude never enjoy; hardly
knowing of their existence, though walking for life
among them. And, like all others, it is a language
of signs.

We can present it only so far as it has been trans-
lated, which will be enough for our present purpose.
These signs are the *characteristics* by which minerals
are known. They constitute, then, the language
which students of this department of nature have
been for ages enlarging and enriching, by discover-
ing new minerals, and studying with more care those
already known. I need but mention these signs to
have it seen that they tax every sense—draw out the
mind by every avenue—pour in knowledge by every
channel, and thus offer the conditions of rapid, well-
balanced mental development.

These signs are, first, the crystalline form.

And what a brilliant language is here introduced.
We have been delighted with the beauty of its characters, even while unable to translate a single word,
and perhaps even ignorant that they were signs of a
language old as creation, sure as the divine oracles,
and varied as the changing figures of the kaleido-
scope. It sparkles from every grain of sand, glitters
from every well-filled cabinet, and streams forth in
joyous, gushing beams from the " Mountain of
Light." These gems, like the stars, have in all ages

delighted men by their brilliancy; but it is in the study of their angles—the planes of cleavage—and the position of their axes, that the ablest minds have found a life employment, and seen the deepest beauties of the mineral kingdom.

It is interesting to trace the progress of mind verging toward truth—peering into the myriad of crystalline forms—coming nearer and nearer to the true translation—sometimes reading a sentence correctly, without daring to vouch for its truth or to join others to complete the story, until Haüy, by the fortunate crushing of a crystal, found in its broken fragments the primitive form, the first intelligible word in this hitherto unknown language. Minds that had been groping in darkness now saw light. Then was called in the power of Mathematics, that ever-ready instrument of progress in science. Whole volumes were filled with geometrical problems relating to this department of Nature. But the wonder is, that in the varied forms into which she molds the outer surface, as if to hide and protect from mortal eye her secret charm, the primitive form within, men should have looked beneath the cunning disguise so as to discover the thirteen fundamental forms from which all others can be derived, and of which they are modifications. As an example of these modifications, we need mention but a single substance—Calcareous Spar, of which Count Brunnon described seven hundred derivative forms. As an aid to mathematical research in reducing these multifarious forms, the light was made to flash the angles by Wollaston's goniometer; and when the forms were determined, the same ray searched many of these crystals through, and by the delicate test of its own polarity, acknowledged the truth of the mathematical deductions. Here, then, on a single characteristic of the mineral kingdom, the crystalline form, have been drawn out the best thoughts of a multitude of laborers—among them De Lisle and Haüy, Phillips and Wollaston. There is no road to the full richness of the mineral kingdom but the one they have opened. To follow that track, even where they have thrown light upon every place of darkness, and placed finger-boards at every doubtful corner, is one of the most severe and accurate processes of discipline to which the mind can be subjected. It is the study of Geometry in material forms—it is the discovery of truth amid a thousand sources of error. The consciousness of being able in such investigations to walk on to truth without failure, in spite of **all disturbing causes, is one of the** most essential requisites to encourage the student, and prepare his mind for original investigations, and the source of the highest pleasure to those possessing it.

I have dwelt at length on this one sign of the mineral language, **because its relation** to intellect has appeared in all the progress of **this science.** It has elicited more thought, its discovery is a greater triumph **of** mind, and it still taxes the higher intellectual **powers more than all other** characteristics of the mineral kingdom combined. Yet they, obvious as they are, have their use—they are the easy part which is first learned ; but when carefully studied, as by some of the old mineralogists, like Werner, they are wonderful in **the accuracy of their results.** The more **important are:** Lustre, with all its various play of light and degree of intensity—Color, with all **its** possible hues—the Degree of Transparency—Refraction and Phosphorescence—Electricity and Mag-

netism—Specific Gravity in all its possible varia-
tions—Hardness of every degree—the **State of Ag-
gregation**—the **Surface, when broken, or scratched,
or reduced to** powder—the Taste and Odor; and if
allowed to step beyond the pure Natural History
properties, we might add the numberless changes
produced by Chemical Reagents. To enumerate the
gradations of these various characteristics would
burden the memory to little profit. **It is seen at a**
glance that they tax every sense. Determinations
depend upon shades of difference so slight, that no
language can describe them; but they are **read in**
the mineral, by that keenness of the senses, which
they always acquire when rightly exercised. Thus
then, in the mineral kingdom alone, we have a language that is never doubtful in its meaning, to the
experienced, but a language to be learned and read
only by the **constant** use of every sense, **and the**
keenest activity of the reasoning powers. What
subject, then, among all the studies of a liberal
education, gives better conditions of mental growth
and activity than this lowest department of nature?

But we must pass to the Kingdom of Life. We

lose here our geometrical forms, bounded by planes and lines, but we have the unfolding of a new force, that uses chiefly four elements, and molds them into more forms than are known to the whole mineral kingdom. The vital force gives relations and developments entirely unlike those in the lower department, not even suggested by any thing found there, —as the relation of parent and offspring,—by which matter is molded into a continued series of identical forms, by a force, not in it, but above it—the development of vegetable and animal growth, in which the perfection and beauty depends upon the constant change of matter, while in the crystal they depend upon its permanence. We have not here stepped beyond the limit of mathematical law, but it is obscured by more deviations than in the most complicated crystal. What myriad forms start up on every side! There is the plant of a single cell, cradled in the northern snow, his kindred lurking in every pool—the Fungus, scavenger among plants, feeding on decaying fiber—the Lichen and Moss, picturing the broad rock with fairy groves and rings —the Grasses, wearing their carpets of green, and

yielding their riches in almost every portion of the
earth—the Fir, and lowly Birch, and Willow,
braving the mountain storms, or creeping almost to
eternal snows—the Pine, whispering its sad moan-
ings in dark and gloomy forests—the Oak, spreading
its arms in strength—the Orange and Citron, loading
the air with perfume—the broad Palm, lifting its
feathery leaves in quiet grandeur to the sky—the
Algæ, binding the ocean with one eternal fringe of
rich and varied hues. Mingled among all these are
thousands of other objects, that make up every
landscape, as rich in product, as curious in structure, and as varied in form. And these all are
ministering to a higher form of life—the animal
kingdom, that, starting from the plant to an opposite
polarity, by a gradation so nice that we can not
draw the dividing line, bursts into a wealth of forms
with sensitive life; ending in man, endowed with
thought and reason, able to understand this chain of
beings, as he is their appointed lord, and their connecting link with the Maker of them all.

Among these we know the Polyp, that with
radiate masonry builds its walls and mounds strong

enough to shut back the ocean, and broad enough
for nations to dwell upon. The waters teem with
Fishes and Shells—the air with Birds and Insects—
the fields and forests with their higher tribes—the
rocks with the casts and figures of those which have
passed away. We have more than one hundred
thousand species of plants; more than two hundred
and fifty thousand in the animal kingdom, besides
the multitude belonging to geologic time. A single
species is sometimes represented by more than one
thousand distinct forms, known as varieties. It is
in this field, among these countless hosts of the
kingdom of life, that the human mind has made
some of its greatest triumphs. This is a matter of
history; but the vastness of the work and the power
of mind required, and the growth of mind marked
by the progress of succeeding generations, can be
fully understood only by those who linger in this
higher portion of the temple of nature till they see
the objects as grouped by the great masters. It
may be said they have only discovered the plan and
the grouping which nature had already made. The
question is not altered. Nature never arranges.

She does indeed put her symbolic language on every stone in her temple. But though the building is perfect to the eye of the great Architect, it is a perfection of *relation*, and not of *position*. It seems chaos to man until that relation is perceived, as it existed in the divine Mind, and is manifested in his works. The blocks are scattered where they were fashioned by the Creator—on every continent, the islands of the sea, and beneath the waters. Their true place is written in their structure: it is repeated in every change, from the unfolding of the germ to the perfect being. But it is the gathering up of these scattered fragments, so that they shall be perfect to man, as they formed a perfect whole to the Omnipresent eye in the first creation—it is this entering in to the thought of God by the army of naturalists, that is the great triumph of intellect. And that this has been done, in the main, in the present natural systems of classification, we have no more doubt than we have that the sun is the center of the solar system, and that the true order of the planets is now known. It is this search, this gradual unfolding of the great Master's thought, that has

quickened the senses and strengthened the powers
of Aristotle, Linnæus, and Cuvier, and the long
list of the dead and living naturalists almost equally
worthy of mention. The record of single struggles
and single triumphs, had we time to recount them,
would prove to us the intensity of thought, the
taxing of the senses, and the broad generalizations
through which each of the great naturalists has
passed,—each being in some points successful, and
in others at fault, because life was not long enough
to read every sign correctly, or because he attempted to form an arch from the materials at hand,
while the key-stone perhaps was fashioned on another continent, reserved as a discovery for some
more fortunate workman. This language of signs,
by which they are compelled to carry on their work,
is the same in kind as already referred to in the
mineral kingdom, but with rhetorical figures and a
more hidden meaning. No other study has demanded of men such bodily toil and exposure. No
worldly good but gold, has ever sent men on such
long and perilous journeys. It so enchains the mind
that ease is forgotten, and money despised except as

a means: it is not valued for a moment against progress in this pursuit.

Agassiz expressed the feeling of every true naturalist, when he said "he could not spend his time making money." Linnæus not only roused his mind and body to the work, so that weariness and disease were almost forgotten, but his pupils were fired with an enthusiasm which sent them round the world, to gather for their teacher and themselves new lines in this book of nature.

There is one department, embracing the whole range of Natural History, whose most brilliant triumphs were reserved for our day, and where the human mind has yet its grandest problems to solve in the material world. Slowly from the mountain and the valley did light break in upon the mind, and the great truth become established, that in the bosom of the earth, that volume of stony leaves, there were strange inscriptions, the record of unnumbered nations; that her true history was written there, and that in this apparent chaos there was perfect order.

The student of antiquities has no lexicon for read-

4

ing the strange inscriptions on the bricks and slabs
of those ancient buried cities. Their engravers,
and those who wrote and spoke the languages, are
gone: not a single letter will ever be added to
those already written; from them alone, unchang-
ing and unchangeable, must a key be found by
which the world can unlock their meaning. Not so
of the inscriptions in the rocks of the earth. The
language engraven there, God is repeating every
year in the sunshine and storm, and in the varied
forms of animals and plants that live and die. This
language the students of nature already knew. As
they opened the leaves of stone, the forms were
strange indeed and antiquated, like the characters in
the old black-letter volumes of our libraries, but the
language was still the same,—it had been the mother
tongue of naturalists for generations. The intellectual triumphs in this field are too recent to need
mention here. The ablest leaders have still their
armor on. But for fifty years there has been no
such field of thought as Geology—no study to which
the universal mind has so turned—none that has
dispelled more prejudice—none that has thrown up

such a background where the thought can rest, or run back through the ages—and there is none that gives more strength of mind by its pursuit.

We have thus far referred to the struggles of mind in unfolding the plan of nature; but has the mission of Natural History been accomplished in its influence upon the great men who have passed away, or is its effect upon mind but beginning to manifest its power? The men already mentioned would have been great in any pursuit. They were lights, though doubtless having greater brilliancy from their peculiar study. Are their works still to quicken and strengthen the mental powers of those who are to come after them; or has the work been done once for all, and is there nothing left for us but to admire the deeds of those giants, without drinking in strength from the same fountain that gave mental vigor to them?

If we mistake not, Natural History is but in the morning twilight of its day of influence. Cast the eye along the shelves of any well-filled library, and see the volumes that have been written to record the labors thus far accomplished. There are Pliny, and Linnæus, and Kirby, and Audubon, and **Lyell**, and Murchison, and Agassiz, and others, the titles of whose books would fill volumes. In what department will **you** find deeper problems for thought, or more attractive subjects for every period of life? We might go further, and say that no class of books is more eagerly sought for, or more generally studied. For the man of general intelligence, or for the scholar, the literature of Natural History is unsurpassed. **What more charming** descriptions than in Audubon and Wilson? What more inspiring than the works of Miller? What authors require **deeper** thought and the exercise of **higher** mental powers than the writers on classification? What works encourage more self-reliance and boldness of views than those of the late geologists?

But the important relation of Natural History to intellect as an educating power, is apparent from **its** modes of investigation—from the objects **it presents** —from the powers it exercises—from the accuracy of its processes and the grandeur of its results.

It calls men to the field, and teaches them to treat of real things, and not of mere names, "terms of ignorance and of superficial contemplation," as Lord Bacon calls them. It thus joins action of mind and body ; gives vigor to the former by its pleasant contrast to mere book-studies, and by giving tone and strength to the latter. Its study is the true method of economizing time in education, for when other books must be closed the book of nature is open; and its subjects of thought meet the eye in our strolls of pleasure, in our hurried walks, and as we rest by the wayside. The swiftness of the car is hardly able to confuse their clustering forms along the way. Our knowledge thus grows in odd moments; and a large portion of life is saved from waste, and made like flower-beds in nooks and borders of gardens, more beautiful because found in places so often neglected.

We shall find no spot on this earth where there is not some alcove of nature's library, with volumes enough to employ us for life. The investigations are always original. The species may be described in the book in our hand, but the particular individual which we are to examine is still to be studied in every characteristic. The description must be seen

4*

to apply; and this, in the ever-varying forms of life, can be done by no mechanical process; it must be by an effort of the mind, apprehending at the moment the entire combination of properties and relations. The first step in wrong theorizing is checked by reference to the real thing, as the calculated distances and angles of the engineer are tested by measurement of the base-line. It thus differs from pure metaphysical investigations, by bringing into constant action the perceptive faculties, as a check to groundless speculations. While Mathematics forces the mind along a given course by the iron rail of necessity, in the relations of geometric figures and algebraic symbols, Natural History compels the mind to direct itself. It must here discover the track, before it can move, and keep itself in place, not by the iron flanges of the car-wheel, but by the quick eye and accurate balancing of the equilibrist. While, then, it allows freedom of movement, it demands accuracy, and corrects error by its constant tests. It does not consist in the dreams of any master's mind, who pities our want of rational insight when we can not understand him, or, understanding him, fail to appreciate him; but it deals with things that have an outward existence, objects that can be perceived and studied by all blessed with five senses. They can be collected in cabinets, so that we may examine the same plant which Linnæus described—the same bone that Cuvier studied.

Natural History demands high qualifications in other departments of education, and constantly increases our knowledge of kindred studies in amount and accuracy, by bringing them into daily use. In the nomenclature, there is needed an intimate acquaintance with the power of words and the laws of their combinations. In considering the geologic forces, the laws of form and position of parts, we gain a clear comprehension only by the aid of Mathematics. In the higher problems of classification, there is a field for metaphysical speculation, applied to no imaginary creations nor abstract terms, but to material forms. The delicate tests of Chemistry, and the almost magic power of Optics, are in constant requisition. Men have become naturalists, it is true, though they neglected other studies; but

such of them as became distinguished succeeded in
spite of their mistake; and in this respect they are
no more to be followed by the student, than the mistakes of Franklin's boyhood are to be copied because he became a statesman and philosopher.

There is no tiring amid the variety of the objects
which Natural History presents, and they cannot be
exhausted. The land and water still abound in un-
studied forms, and the scalpel and microscope reveal
new wonders in those that are old. They are generally beautiful in themselves, always beautiful in
their relations, so that the mind is constantly re-
lieved by new points of interest, and thus dwells
upon them without weariness. They daily meet the
eye, and invite us to review. Other studies may be
forgotten because the books are closed and gathering
dust on the shelves, but the flowers and the trees
can not thus be put away. They press themselves
upon the attention every day, and the insects and
the birds will have a hearing. If the cold of winter
drive them away for a season, they make up for the
loss when they return in the spring, filling every tree
and bush with their melody. Whoever heard of a

naturalist forgetting or losing his interest in his studies? Those who have contented themselves with learning a catalogue of hard names, supposing this to be Natural History because it often passes for it, must expect to lose this, with most other knowledge held by memory alone. Men may name whole cabinets, and have no more claim to be called naturalists, than a man who has simply learned a hundred words from a Greek Lexicon to be called a linguist. Such knowledge costs more than it is worth to keep it. The best thing that can be said of it is, that it seldom troubles its possessor long. But he who has once seen the true plan and relationship of natural objects is a Naturalist, though walking among animals and plants that have never yet received a name; and the knowledge of that plan and relationship can never be forgotten, but will be increased by every new object which meets his eye.

When the mind would mark the nice distinctions drawn by Nature, she must call to her aid every sense. She must read the cells in the bone and the glimmering lines of the scale—the veining of the leaf and the angle of the crystal. By being thus

drafted to constant labor, the senses are so changed
in degree that they seem almost new in kind. Distinctions are marked, threads of truth gathered up,
which unpractised senses can not perceive, nor minds
untrained to like studies appreciate.

This accounts for the common undervaluing of
the most important labors of the Naturalist. What
need of blinding one's self in studying microscopic
organisms and the mere impressions in the rocks?
Because they are links in the chain—tints in the
grand picture. As well might the linguist neglect
the breathings and accents of his Greek language,
the astronomer his fractions of a second, as the naturalist these minute and seemingly useless objects.
As well might men sneer at the painter for giving
those fine touches that mark the works of masters,
or at the sculptor, as his chisel brings out, by its
fine cutting, the desired expression, as at the natur
alist when studying these minute shadings on the
great canvas of nature. It is by these intershadings
alone that the parts are seen to form an harmonious
whole, in the contemplation of which the mind is
both delighted and truly educated.

In educating the mind, accuracy is one of the most desirable traits to be developed. Volumes have been written that are worthless for lack of this element. We feel no safety in consulting them. Fine intellectual powers have yielded no valuable results in the labors of a lifetime, because not directed by habits of accuracy in every undertaking. A mind that rests on suppositions is never to be trusted by others, and can never satisfy its possessor, if he have keenness enough to understand his own defect. We all feel the power of this in every pursuit. We wish to trust life and fortune with the accurate men. And if we would give to those whom we educate the highest mental culture, they must be taught to scan every relation, and mark the minutest bearing of every subject brought under their consideration. This may be a natural gift to a favored few, but to the majority of men it comes only by careful training. In every branch of study chosen for its educating power, this characteristic of securing accuracy in every mental process is considered of the highest importance. And from the whole range of studies in the most liberal course, we challenge the

selection of one that demands accuracy, and secures
it more fully, than Natural History, as now studied.
Look at the Botanist, as he marks every hair, and
line, and cell, when with microscopic power he looks
into the secret laboratory of life, and traces the joining of the tissues and the structure of the minutest
organ. And in this respect the Zoologist is wholly
his equal. He studies thousands of microscopic
forms—the wavy line of the scale, and the cell of the
bone—the cells, and lines, and tissues of the egg,
from the first crimson tinge of life, till every change
has been completed. The power and accuracy which
this gives are seen in the restored forms of vegetable
and animal life from the scattered fragments in the
rocks. This power and this habit, as a part of education, appear in every vocation of life.

Another requirement of a study is, that it shall
give broad views, and make men liberal towards
other pursuits. Accuracy is dearly bought if it nar-
rows the mind, so that it can see no good in any
thing beyond its own particular province. Natural
History calls into daily requisition almost all other
departments of human knowledge. It does this in

so marked a degree, that their true place can never be lost sight of, nor their value underrated.

In the grandeur of its results, Geology is, according to Sir John Herschel himself, second only to his own favorite study, Astronomy. Humboldt, whose range of knowledge is certainly equal to that of any man who ever lived, and knows well what studies are requisite to breadth and completeness of view, has placed the study of an humbler branch of Natural History on equality with the sublime study of the heavens, for securing accuracy and intellectual power.

"The Astronomer," says he, "who by the aid of the heliometer, or a double-refracting prism, determines the diameter of planetary bodies, who measures patiently, year after year, the meridian altitude and the relative distances of stars, or who seeks a telescopic comet in a group of nebulæ, does not feel his imagination more excited—*and this is the very guarantee of the precision of his labors*—than the botanist who counts the divisions of the calyx, or the number of stamens in a flower, or examines the connected or the separate teeth of the peristoma

surrounding the capsule of a moss. Yet the multi-
plied angular measurements on the one hand, and
the detail of organic relations on the other, *alike* aid
in preparing the way for the attainment of higher
views of the laws of the universe."

It is with such views of the benefits of Natural
History that we would have its study entered upon
by the young. It may not bring money to them,
but it will open new sources of pleasure. Nature
will become an exhaustless volume, read with delight; and not simply a series of pictures which they
can admire indeed, but only as children do their
primers, without a thought of the story, or at least
without the ability to read it. Thousands have ad-
mired the beauties of the moss covering the earth
with an elastic carpet of green; but how is that
beauty heightened to a Humboldt, when he sees in
the microscopic points in its nodding capsule a
new note in the harmony of the universe!

If we look then at the long catalogue of honored
names, whose whole lives have been given to the
study of Natural History—if we look at the volumes and cabinets which now record their labors

—if we look at the power of this study to develop the perceptive faculties—if we look at the accuracy of its processes, and the grandeur of its results—and above all, if we look at these varied forms, as the material expression of the thought of God—it comes to us with a force that needs no special plea to sustain it, that Natural History is deserving all the labor men have ever bestowed upon it, as a means of training the intellectual powers, and as one of the most delightful fields for their exercise.

LECTURE II.

NATURAL HISTORY AS RELATED TO TASTE.

ALL created things are a series of dependencies. They are more than a simple series; for, if studied in groups, like the stones of an arch, each group is found to have not only its own conditions of existence, but to be conditional for another. This is certainly true of all material forms of organic or inorganic combinations. So far-reaching is this idea of relative dependence, that we can not by reasoning reach an absolute in time, space, matter, or force. Go far as we may in either direction, the mind still seeks for a more remote cause, a still lower condition. But by allowing the mind to rest on a single view and to consider a single condition, this view or condition becomes magnified; and though it can never become to us the absolute, it may assume an undue importance, and seem to us to be the very atlas upon which the world rests. All members of the series are stones in a perfect arch. The stone we delight in, we see to be necessary to the very existence of the structure, and we forget that this is true of every other in the sweep of the all-embracing curve. This tendency to consider every object of our interest or study as a condition for other good, rather than as itself equally depending upon others, is seen in every pursuit; and probably no man is so liberal in his education as to be entirely free from this tendency. The study of nature is thus judged of and directed according to the stand-point of the observer. Each one has his own measure of utility, and nature is to him valuable as she seems to expand when he applies this test. The mere man of business sees in money the hope of the world—the mainspring of progress, and the price of every thing desirable. What are the laws of Mechanics to him, but that his warehouses may be strong and his machinery fitted for its work? What use of Astronomy, but for the guiding of ships, to shorten the passage and reduce the insurance? What good in Natural History, but that the earth may be made to yield more abundant products from her soil, unlock her mines of coal, and become a grand specie-paying

5*

bank, always discounting freely, but never demand-
ing pay? The mere student inquires its relation to
the Intellect, and his scale of worth measures its
power of exercising and developing the faculties of
the mind. The artist rises higher still, and, above
all notions of wealth, above the pure conceptions of
intellect, he ranks the emotional nature—the love
and enjoyment of the beautiful for its own sake.
Nature to him has value as the Cosmos, revealing a
mind, and speaking to the mind, in its varied language of order, proportion, and grandeur; thus
awakening the emotions of beauty and sublimity.
These may, indeed, arise from very general views,
hardly to be ranked as the study of Natural History;
it is, however, the study of each particular part that
brings out the keener enjoyment of the soul, as the
fine tones in music add deliciousness and richness to
the harmony. But, rising higher still than all per-
ceptions of material beauty, all enjoyment from the
possible combinations of matter, is the spiritual
nature and sense of moral beauty. To this all other
sources of happiness must be inferior, if not conditional, for it is in this direction that man approaches

nearest the Maker of all, in whose likeness he was formed. To those considering alone this higher spiritual and moral nature of man, the material world becomes simply a divine revelation. This is to them not only its highest worth, but, in contrast, all other purposes are underrated, if not despised. It may be well for the world that men should thus view truth from different stand-points, and become so enamored with a single side as to gaze at it for life. But it is never well for the individual. A thousand minds, fixed on a thousand different points of the same object, must in the aggregate learn more than one possibly could by passing from point to point. A thousand men, content thus to rivet their attention till the smallest object filled their entire vision, would make greater progress for the world, but it would be at the sacrifice of individual advancement. The broader the views, the more correct. And that mind is alone well balanced that can glance through the whole range of relations, and give to every faculty of mind and department of nature its true position. For the arch is beautiful and perfect only when the key-stone rests in its

highest point. In the study of Natural History, its
entire relation to man is to be considered. The
Intellect, as we have already shown, finds here a
soil adapted to its growth. Like a sturdy tree, it
may here strike its roots deep, and send up the
heavy trunk, and broad branches, and load them
with golden fruits. Here, too, Taste may flourish
under the same favoring influence, as pure intellectual culture; like the vine or prairie-rose upon the
oak, twining in graceful folds, and spreading over
the broad, firm branches of intellectual growth an
eternal adorning of indescribable beauty.

It is on the relation of Natural History to Taste
that I wish to speak at this time.

There is in man a love of the beautiful. And by
the beautiful we mean that which delights by simple contemplation—that which we admire without
the thought of utility, and without the ability,
perhaps, to explain the cause of our admiration.
The emotions excited by beauty and grandeur may
be pronounced simple or complex, in our analysis of
the emotional nature, but "they are," says Allison,
"distinguishable from every other pleasure of our

nature." "The qualities that produce these emotions are to be found in almost every class of human knowledge, and the emotions themselves afford one of the most extensive sources of human delight. They occur to us amid every variety of external scenery, and among many diversities of disposition and affection in the mind of man.

"The most pleasing arts of human invention are altogether directed to their pursuit. And even the necessary arts are exalted into dignity by the genius that can unite beauty with use. From the earliest period of society to its last stage of improvement, they afford an innocent and elegant amusement to private life, at the same time that they increase the splendor of national character; and in the progress of nations, as well as individuals, while they attract attention from the pleasures they bestow, they serve to exalt the human mind from the corporeal to intellectual pursuits."

The faculty or constitution of our minds by which we perceive these qualities, and enjoy these emotions of beauty and sublimity, is Taste. It is itself a plant of beauty in the garden of mind, but crushed and despised in the hurry of this **utilitarian age.** It has too often been neglected by the scholar, and mourned over as a vile weed of depravity by the Christian.

The pleasures **of this faculty are to the** individual ever **fresh and delicious.** But while **every emotion** of beauty thrills the soul with delight, it rolls **in hurrying ripples, and** leaves only for its possessor **conscious evidence of** its value and elevating power. There is, therefore, **in respect to this** faculty, an individual growth and revenue **of pleasure,** which no one can calculate for another. **We can take no** inventory of these higher riches, though **in respect to them, men undoubtedly differ more** than they do in material wealth. But there is for the race an outward expression of the power of Taste, and a permanent record of its progress, in the Fine Arts. If they are not the creations of Taste, they are, some of them at least, **the creations of Genius to supply** her demand; **and the highest aim of** Genius is but **to** receive her approbation. To the bidding of this goddess he yields the tribute of all his powers, and plumes his wings for his highest flights with the devotion of a knight in the days of chivalry. By her demands, some arts are made to minister more to her gratification than to bodily wants, and these we raise from the rank of the simply *useful* to the *fine*.

The intellect simply demands of language that it express the thought with clearness and precision,— but at the bidding of Taste, Genius weaves it into the gorgeous web of poetry, gleaming with threads of gold, and covered with the most brilliant hues that fancy can paint—the most pleasing forms that imagination can combine. And even the language of common thought it has adorned with gems and flowers. Her mandate has changed music from the harsh and grating sounds of savage instruments to the richness of cathedral organs, and the magic bow of Ole Bull. Painting and Sculpture, for her delight, have enlisted the pencil of Raphael, and the chisel of Michael Angelo. For her, Architecture raises the fluted column, places the molding, and spans the arch. The landscape, for the thorn brings up the fir-tree, and for the brier the myrtle-tree, and becomes a place of enchanting views under the genius of a Downing.

These arts, in their lowest forms, alone minister
directly to the physical wants of man. It is in
obedience to the commands of this higher power of
mind, to which their beauty is addressed, that
Genius has brought them to their present degree of
perfection. Whatever they have of beauty or
sublimity, has been given to them to meet the
demands of Taste. To them we may look to see
what drafts have been made from Natural History.
While Genius has explored the field of thought, his
work was not done till he added to these immaterial
beauties the crystal and flower and forms of sensitive life—beautiful in themselves, and symbolic of
those higher beauties beyond the reach of the
senses. By marking the road along which Taste
has led her votaries, we may learn where her
onward path must lie, and how far Natural History
can furnish material for building or adorning the
beautiful structures which she demands.

The adorning of thought, by language not needed
for its mere expression—that portion created at the
demand of Taste—is one of the highest works of
genius, and for this alone natural objects would be

worthy of more study than they receive. Not only
do they themselves awaken every emotion related
to Taste, but it is by them alone that we express
the higher moral beauties and relations of thought
which it is in our power to conceive. Even God
himself gives the precepts of his revealed will, and
sets forth the glories of his Church, by the use of
these very objects; and, so far as we can see, there
was no other way in which it could so well be done,
if at all. Glance for a moment at your favorite
authors,—the poet, whose sweet song charms and
gives enjoyment by its very refining power; the
orator, whose words enchained every listener with
their beauty,—and see how much they are indebted
to symbols drawn from nature. Their words may
be joined by the rules of grammar and logic—they
may convince the Intellect by the force of the
reasoning, they may arouse the will by the plea of
interest—but when they would charm with beauty,
they must reach forth for the gems and flowers of
Nature.

There is indeed much borrowed from Nature to
beautify language, that is not strictly Natural His-

6

tory. The stars glitter in literature almost as they
do in the heavens. The bands of Orion and the
sweet influence of the Pleiades, and all the famous
constellations, have beautified almost every language. To these the Naturalist can lay no claim.
And it may be said that the writers who borrow
their illustrations most largely and successfully
from the objects of Nature, are not Natural Historians. They may not study books to learn those
natural objects they have never seen—they may be
ignorant of the terms of Linnæus, and the divisions
of Jussieu. They may not be able to give a single
scientific name, and yet every writer that pleases us
most, looks with the eye of science, and describes
with the accuracy of a Naturalist. Their vivid and
minute descriptions show the skill and strength of
the observing power. The effect of this is seen
even in the savage, before brutalized by the white
man's vices. He puts to blush the best-trained observer of the schools, and marks with the naked eye
nice distinctions, which even the microscope can
hardly reveal to some of us. These forms of Nature
give him not only the graceful model of his canoe,

and the delicate tracery of bead-work without a pattern, but also the symbols of his expressive language. His words are of leaves for number, the
rose and violet for beauty, the eagle for swiftness,
the fawn for gentleness, and the snake for stealth.
There is beauty in his language, and it is borrowed
from natural objects, and every thing written respecting him draws necessarily its beauty from the
same source.

When the poet would sing of the Indian's legends
and traditions, he repeats them "as he heard them
from the lips of Nawadaha, as he found them

> "In the bird's-nest of the forest,
> In the lodges of the beaver,
> In the hoof-prints of the bison,
> In the eyry of the eagle."

The Indian's allegory of Winter and Spring beautifully illustrates their use of the bright images of
Nature.

> "When I shake my hoary tresses,
> Said the old man, darkly frowning,
> All the land with snow is cover'd,
> All the leaves from all the branches

Fall and fade and die and wither;
For I breathe, and lo, they are not!
From the waters and the marshes
Rise the wild-goose and the heron—
Fly away to distant regions."

* * * * * *

"When I shake my flowing ringlets,
Said the young man, softly laughing,
Showers of rain fall warm and welcome,
Plants lift up their heads rejoicing;
Back unto their lakes and marshes
Come the wild-goose and the heron,
Homeward shoots the arrowy swallow,
Sing the blue-bird and the robin.
And wher'er my footsteps wander,
All the meadows wave with blossoms,
All the woodlands ring with music,
All the trees are dark with foliage.

"Then the old man's tongue was speechless,
And the air grew warm and pleasant,
And upon the wigwam sweetly
Sang the blue-bird and the robin;
And the streams began to murmur,
And a scent of growing grasses
Through the lodge was gently wafted;
And Segwun, the youthful stranger,

More distinctly in the daylight
Saw the icy face before him—
It was Peboan, the Winter.
From his eyes the tears were flowing,
As from melting lakes the streamlets,
And his body shrunk and dwindled
As the shouting sun ascended,
Till into the air it faded,
Till into the ground it vanish'd,
And the young man saw before him,
On the hearthstone of the wigwam,
Where the fire had smoked and smolder'd,
Saw the earliest flower of Spring-time,
Saw the beauty of the Spring-time,
Saw the Miskodeed in blossom."

Along the whole stream of ancient song, the objects of Natural History are set in thick and sweet profusion—not gathered into clusters, but adorning the richness of the poetic imagery as flowers deck the meadows; and the soft numbers seem to flow like pearly streams reflecting the nodding verdure on their grassy banks. How beautifully are they braided into song, as a chaplet for the tomb of the Grecian poet!

"Ye evergreens, around the tomb
  Of Sophocles, your osiers braid,
  And, ivy, spread thy pensive gloom,
  To form above the bard a shade.

"And intertwine the blushing rose,
  And gentle vine your leaves among,
  Thus, gemm'd with beauties, shall your **boughs**
  Prove emblems of his graceful song."

In the Pastorals of Theocritus, and the morning
song of Moschus, the flowers bloom and the trees
whisper. What lent the charm to much of Virgil's
poetry—the Bucolics and the Georgics? His
sweetest strains are mingled with the hum of bees,
and the song of birds.

"Behold! yon bordering fence of sallow trees
  Is fraught with flowers, the flowers are fraught with bees.
  The busy bees, with a soft murmuring strain,
  Invite to gentle sleep the laboring swain;
  While from the leafy elm **the** turtle-dove
  Tells in soft notes the story of its love."

Thus through that wonderful poem, written at the
command of his sovereign, has he presented a pic-

ture of nature, such as the Naturalist delights to contemplate. Not indeed accurate in all respects—we have in many points the ignorance of the times, and absurd theories; but mingled with this, the accurate description, and the fresh painting of natural objects, which made the work a blessing to Italy and the delight of every age.

But poems in our own language are not only quite equal in this respect to the Greek and Latin, but surpass them. Thompson sings of the seasons; but they are the grand moving panorama, that would be blank canvas, but for the objects of nature which follow in quick succession. He paints with a master's hand, and the charm that envelops the whole is the picturing, so true to nature that you seem in the mine where crystals shine, and by brooks where the flowers blossom. In his tribute to the sun, the gems seem to glisten as though set in a coronet of beauty.

> " The lively diamond drinks thy purest rays,
> Collected light, compact, ...
> At thee the ruby lights its deepening glow,
> And with a waving radiance inward flames;

From thee the sapphire, solid ether, takes
Its hue cerulean ; and of evening tinct
The purple-streaming amethyst is thine.
With thy own smile the yellow topaz burns,
Nor deeper verdure dyes the robe of spring,
When first she gives it to the southern gale,
Than the green emerald shows. But all **combined,**
Thick through the whitening opal plays thy beam."

This reminds us at once of the beautiful description of the Russian jewels by Bayard Taylor, whose language seems rich and brilliant as though gilded with the light of the gems it describes.

"The splendor of their tints is a delicious intoxication to the eye. The soul of all the fiery roses of Persia lives in these rubies, the freshness of all velvet sward, whether in Alpine valley or English lawn, in these emeralds ; the bloom of all southern seas in these sapphires, and the essence of a thousand harvest-moons in these necklaces of pearl."

We might thus follow our own poets through this same path, as they not only adorn their language by introducing objects of Natural History, but have so faithfully described the various objects in all its kingdoms, that they teach as well as delight us. In

Bryant's poems the beauties are truly the beauty of Nature. The flowers blossom, and the birds sing. The grove is filled with life, and every object is drawn with a master's pencil, that gives Nature's own form and color to the streak of jet on the violet's lip. To meet the demand of Taste, these sons of genius and of song go forth into Nature's ample field to select their subjects and their illustrations. Heroic verse might flourish in an earlier age, when heroes were demi-gods; but for the beauty of our English verse, we have no more propitious muses than the birds and flowers, no loftier Parnassus than the hill of science.

If we needed higher illustration of the power of natural objects to adorn language and gratify Taste than we have in the poets, we should appeal at once to the Bible. Those most opposed to its teachings have acknowledged its beauty, and this is due mainly to the exquisite use of Natural History objects for illustration. It does indeed draw from every field. But when the emotional nature was to be appealed to, the reference was at once to natural objects, and throughout all its books, the objects of

Natural History are prominent as illustrations of the beauties of religion, and the glories of the Church.

How could the most refined taste be more highly gratified, than by some of these beautiful illustrations of prophecy?

"The wilderness and the solitary place shall be glad for them, and the desert shall rejoice and blossom as the rose."

"The mountains and the hills shall break forth before you into singing, and all the trees of the field shall clap their hands. Instead of the thorn shall come up the fir-tree, and instead of the brier shall come up the myrtle-tree."

We know that it was no mere lover of Nature in the general, but the royal student of Natural History, who knew plants, from the cedar of Lebanon to the Hyssop in the wall, who penned that picture of nature which never can be surpassed for its beauty.

" For lo, the winter is passed, the rain is over and gone, the flowers appear on the earth, the time of the singing of birds has come, and the voice of the turtle is heard in our land; the fig-tree putteth forth her green figs, and the vines with the tender grapes give a good smell."

The power and beauty of these same objects appear in the Saviour's teachings. The fig and the olive, the sparrow, and the lily of the field, give a peculiar force and beauty to the great truths they were used to illustrate.

The glories of the holy city in the Apocalyptic vision could only be set forth in the symbols of gems. Its foundations were of sapphire and emerald, of topaz and amethyst. And every several gate was of one pearl.

Thus, then, in all adorning of common language, in the beauty of poetry, and in the vivid pictures of divine inspiration, the sweetest note that strikes the ear comes from the landscape, the brightest picture is the landscape itself. All that Taste has ever demanded for her gratification, Genius has here found, and as God is the author of both nature and mind, here among the crystals, flowers, and sensitive life, must the emotional nature of man find its highest earthly gratification.

In painting and sculpture, the human mind is striving for the same that appears in poetry, and the adorning of common language. That love of the beautiful must not only be gratified with descriptions upon which the thought can dwell, but we would look into the minds of others and see the pictures into which imagination weaves these objects for them. It is only as they can present to the the eye, by the pencil and chisel, the subjects of their thoughts, that we can compare our imaginary scenes with theirs, and learn what different emotions the same words and the same objects awaken in different minds.

As nature is the storehouse from which writers draw, and the pattern according to which they must work, so must this also be true of the painter and sculptor, who would trace upon the canvas, and chisel from cold marble, figures that shall glow forever with the warm expression of life.

There is a mathematical law of development, and a constancy of expression in the minute markings of species that nature never omits, which can never be neglected by the artist, if he would meet the demands of that true taste which delights in the truthfulness of works of art, rather than in the glare of colors, or the grotesque in form.

Poetry, painting, and sculpture have moved on together in all ages. "The whole compass of ancient poetry was in fact reshaped in the marble of the Grecian sculptors, and delineated anew on the canvas of the painters." Perhaps this union of the three is not so strongly marked in our time, but though diverging more, they are still like triple stars of complementary colors, all forming one system, and all needed for the expression of the emotions of Taste, and each moving in an orbit varied by the others. While one is in the heathen heaven, among the gods and goddesses, the others are there also, and when one returns to earth, the others bear her company. On the canvas and in the marble, are the sensible expressions which poetry created,—though the poet's brain, and the painter's and sculptor's cunning, have sometimes been the possession of a single man. He is the true genius, and we know what in his creations

7

gives us most delight; it is the truthfulness of
nature which they present—a truthfulness becoming
more apparent as they are longer studied. We do
not expect in his productions the serrate mountains
of granite, where nature has covered the hills with
the smooth belts of slate and softer stone.

We may not be able to point out every fault in a
work of art—from our defective education we may
even praise such works when faulty ; but it is a law
as established as the courses of the stars, that works
of art live only as they have the beauty and truth
which accurate study of natural objects can alone
give them.

This is the ground of Ruskin's criticism of the
famous statue of Laöcoon. We may remember
that all the circumstances were out of the ordinary
course, and thus be carried along by the power of
the poet and the skill of the sculptor; but in
ordinary pieces, snakes must not feed like wolves,
but, true to their nature, only crush by their tight-
ening folds.

The need of accurate study of nature is proved by
the practice of the best masters. The painter and

sculptor study every bone and muscle with the accuracy of the naturalist or the professional anatomist. The statue which seems like an enchanted form of Arabian tales, ready to start into life at the first blast of the trumpet, and the charger with expanded nostril and rearing form, so life-like that he seems bounding from his granite pedestal, were no creations of mere casual study. But days and weeks, the points of living expression were fixed by the same study that must be given by the Audubons and Agassiz of Natural History.

With painting and sculpture, in ancient times, architecture was intimately connected. Though this relation can never be broken, it is not now so marked. With us, architecture, so far as it relates to adorning and beauty of expression—and in these respects alone can it be denominated a fine art—is intimately connected with landscape gardening. This may not be true of public buildings—they still must borrow their ornaments from ancient patterns; but so far as architecture can be applied to the homes of the people, it is united with gardening, which has been raised from the rank of the useful

arts, to one of the most **effective means of ministering to Taste.**

We might repeat in reference to ancient architecture, **what we have already** said of the necessity of
the study of nature ; for it speaks from the broken
master-pieces **chiseled under the eye, if not by the**
hand of Phidias. The very form and peculiar
ornaments of some of the orders—the acanthus **of
the** Corinthian capital, the points and arches of the
tree-formed Gothic—only have **their full expression,**
and the expression its **full appreciation, from a care-**
ful study of nature.

But for our homes, **we have** exchanged **the forms**
that were **the** offsprings of mythology and supersti-
tious reverence of the gods in high places, for the
rustic beauty of varied forms more pleasing to the
rural deities, which are the only ones our fancy can
still perceive lurking **in our** glens and **among the**
groves yet spared by that avaricious **Vandalism**
which has **stripped of their ornaments so many** hill-
sides.

Home architecture and landscape gardening are
necessary complements of each other—together they

must grow, becoming more beautiful at the demand of Taste; culling every flower, twining every vine as in its own native thicket—inviting even the birds, until home itself shall seem to have sprung from the earth, at the touch of some magician, whose whole soul had drunk in the beauties of the river, plain, and mountain. In this department of the fine arts our country has most to hope, for the poorest man can enjoy it as well as the rich. Money is not wanted, as in the purchase of costly pictures and fine statuary, but nature offers the beauties—all we need is the eye to perceive and the power to combine the materials which she furnishes. These constitute the democratic division of the fine arts, equal to the best, and yet within the reach of all. There are true, elevating, and unfailing sources of enjoyment, which the poorest laborer can enjoy as free as the air of heaven. They are in the field he tills, along the road he travels, in the ocean he navigates; everywhere he looks he might see more beautiful objects than adorn the galleries of the richest nobleman. But to see them he must be taught to observe. He must study every object till

7*

he perceives its beauty, "for be sure it is there." Perhaps some general admirer of nature may imagine he at least has seen and perceived—for there is a vast difference between them—all the possible beauties of nature. He believes this beauty to be found only in the general effect, and not in the single objects as studied by the Natural Historian. Does not the general effect of the picture depend upon single lines? If you think you have by this general survey discovered all the beauties of nature, walk into your own fields with the Mineralogist, and you may see crystals gleam where you never suspected their existence; go with the Botanist, and new flowers will seem to spring up along your path, and new beauties appear in those known to you before; the Entomologist will drag from his lurking place "the beetle, panoplied with gems and gold;" the Ornithologist will point out new birds which have been seeking your acquaintance since childhood.

We have seen this effect in young persons in a course of education. They professed to admire Nature, and to be able to perceive her beauties. Let now the study of Natural History demand of them

accurate and systematic study, and it seems almost
to implant within them a new sense. What exclamations of surprise and admiration break forth
from them in their excursions! What new flowers
they now discover!—they have been treading upon
them unheeding their beauties all their lives. What
strange birds!—they have been flitting above their
heads for twenty summers. And now, by this simple process, there is awakened the power of perceiving and appreciating the beautiful, that seems like
the richness and music of spring compared with the
death of winter. When carried farther, there comes
the power of combining these objects so as to repro-
duce, when we please, the same sweet scenes which
nature plans in some far-off hill or glen. To produce this general effect, it may be thought that
only general notions are needed. This is undoubtedly true, if we refer to the emotion of grandeur, in
producing which magnitude is more powerful than
form. But for the emotion of *beauty*, we must have
these objects arranged as he only can arrange them
who has studied their minutest marking, every form,
and every tint.

This power of combining to produce the effect of
nature, like a simple style of writing, *seems* easy to
all, but is hard to acquire. And one who commences it supposing he shall succeed because he has
been a general admirer of nature, will have occasion
to blush for his mistakes, and will find it hard to be
natural unless he takes long and patient lessons of
the only teacher, Nature herself—fixing his eye
upon every object till its last touch is stamped upon
the mind. Then it can be used, then it is a possession, and "a joy forever." The power which this
study gives is well illustrated in the mounting of
birds, which some think ought to be reckoned
among the fine arts. The learner may become skillful in the manual part. Every feather may be in
its place as pure and unruffled as in life—the eye of
glass may rival the real eye in brilliancy, and still
there is death. One touch from the master's hand,
and you almost start back from the living bird.
The power of life lights the eye and seems to reach
the tip of every feather. Whence came the magic
power? It came from the careful study of the bird,
till every varying change of life was daguerreotyped

in the mind. If the common mind is to be trained
to the love of the beautiful, it must be in the great
gallery of nature, and by gazing like students before the works of the great masters, till every line
and tint are fastened in the mind, and beauty is living in the soul.

Lord Kames tells us that "those who depend for
food on bodily labor are totally void of taste, of
such a taste indeed as can be of use in the fine arts."
He would hardly have written that in our day.
We seem to see Hugh Miller come up from the
hard work of Scotland's stone quarries, with a soul
as noble, a taste as refined, with the highest emotions as keen as he looked away upon the varied
landscape, with the eye of a naturalist and the soul
of a poet, as the wealthiest lord ever possessed when
walking among the works of art that only princely
wealth could purchase. No other language can
equal his own glowing description, as he thus records the experience of his second day as stone
quarryman. "I was as light of heart next morning
as any of my brother workmen. There had been a
smart frost during the night, and the rime lay white

on the grass as we passed onward **through the fields**; but the sun rose in a clear atmosphere, and the day mellowed, as it advanced, into one of those delightful days of early spring, which give so pleasing an earnest **of whatsoever is** mild and genial in the better half of the year. **All the workmen rested** at mid-day, and I went to enjoy my half-hour alone on a mossy knoll in the neighboring wood, which commands through the trees a wide prospect of the bay and the opposite shore. There was not a wrinkle on the water, not a cloud in the sky, and the branches were as moveless in the calm as if they had been traced on canvas. From **a wooded** promontory that stretches half way across the frith, there ascended a thin column of smoke. It rose straight as the line of a plummet for more than a thousand yards, and then on reaching a thinner stratum of air, spread out equally on every **side, like** the foliage of a stately tree. Ben Nevis **rose to** the west, white with the yet unwasted snows **of** winter, and as sharply defined in the clear atmosphere, as if all its sunny slopes and blue retiring hollows had been chiseled in marble." . . . .

"I returned to the quarry, convinced that a very exquisite pleasure may be a very cheap one, and that the busiest employments may afford leisure enough to enjoy it."

There is a growing Taste among our people—it is sad indeed that its growth is so slow—which proves that honest toil does not destroy nor dwarf the capacity of enjoying the beautiful. It can not, however, be fostered by galleries of art, for they are rare among us. It is upon Nature we must depend; and Landscape Gardening, by the genius of Downing, is gathering scenes of tasteful beauty around many a humble home. His works were to America, what the Georgics were for ancient Italy. The vine and the apple, the flower and the hedge, the velvet lawn and stately tree, all that beautifies the landscape, were objects of his care. Through his influence, many places are pleasant to the eye and refining to the taste, which but for him would have remained rugged and neglected.

The homes in cold, rugged New England, in the sunny South, and on the western prairies, will have more beauty, and the children reared there will be

men and women of more refinement, **because Downing** was a lover of Nature.

It is meet that his monument should stand upon
our **national grounds at** Washington, not only because they **were beautified by his hand,** and because
his influence was national, **but that every American**
might read the words he penned while **living, now**
engraven on the stone.

" *The taste of an individual as well as a nation
will be in direct proportion to the profound sensibil-
ity with which he perceives the beautiful in natural
scenery.*"

Thus has Natural **History ever been** the field
where the objects of taste have been gathered in the
greatest abundance, and it must ever be the great
source of the pure and beautiful images which the
progress of the Fine Arts demands.   The cultivation
of Taste is sneered **at** by those who talk wisely of
utility, but its value can not **be over-estimated; and**
its progress must **move on necessarily** with the
study of Nature, especially with that more accurate
study which we denominate Natural History.

The accurate study of this science stores the mind

with images of things formed by God himself; they are, then, so far as art is concerned, "the true and the beautiful." This it accomplishes by educating the senses. It also prepares men to receive and cherish every form of beauty, by carrying the thoughts up to the divine source of all created things, thus developing the higher spiritual nature and purifying the soul. In a polluted soul, no perfect image of beauty can dwell—it can not be formed there; like a distorted mirror, the clearest light may fall upon it and the most beautiful objects may pass before it, but the images formed will be changed in proportion or relation. The beautiful will be reflected as hideously deformed, while the loathsome and horrid may be thrown back distorted into the perfect. But to the mind and soul capable of perceiving, nature offers standards in color, form, relation, and proportion, set by Him who is the author of mind as He is also of the external world, and therefore they must be correct. "He that formed the eye, shall he not see?"—and He that formed the mind, shall He not understand its wants and provide for the demands of Taste as perfectly as He has for every other **want of our being?**

The whole history of the Fine Arts shows that God has here established immutable relations, and those works alone **have stood the** test of time that approach the patterns which **He has given.** The voice of the Most High speaks to the artist as **to** Moses in the building of the Tabernacle—"And look that thou make them after their pattern which was showed **thee in the Mount."**

The study of nature is within the reach of all, and if studied as it ought to be, the many may become judges of the objects **of taste,** rather than the few. **The effect** of this on **the Fine Arts** would be marked. It is said that the most illiterate shopman of Rome is a better judge of pictures and statuary, than those of the most refined education in London, which certainly has many advantages over most American cities. In such a place as Rome, **a poor work of** art could **hardly be produced**—certainly could never be praised for excellence. We have no such means for creating a correct taste for these works, because galleries of art are rare, **and in our hurry** we might not find leisure to study them. But we have a noble and neglected substitute—the beautiful objects of nature, which might delight us even in hours of hardest toil.

In the effects produced by objects of Natural History, we have referred almost exclusively to the emotion of beauty—but they certainly offer for contemplation the grand and sublime. What grander field for the imagination than is offered by the revelations of Geology? The object presented may of itself be insignificant to a common mind, not perhaps perceived, or if noticed it does not awaken a single emotion. How very different the same mark or pebble may become to the student! For him, a single line across the granite of the mountains carries the mind back to the time when Neptune made war against the hills, and hurled against them his whole enginery of waves and ice. A single vein in the rock summons up the scenes of the Plutonic dynasty, whose records are the everlasting hills and the dykes that divide the broken strata. As he unfolds the stony leaves of the earth, a thing of beauty, a single fossil, may tell to his

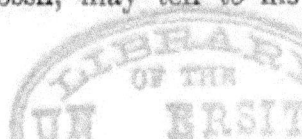

instructed mind a story of grandeur and sublimity
It may repeople the earth with wondrous forms,
pour the oceans upon the sinking land, and move
the hills like watery billows.

The fancy roams through all the beauties and
grandeur of the early earth. If we have never had
the privilege of studying one of Nature's galleries of
ancient art, we can not do better than to hear Buck-
land describe the richness of the Bohemian coal
mines.

"The most elaborate imitations of living foliage
upon the painted ceilings of Italian palaces, bear no
comparison with the beauteous profusion of extinct
vegetable forms with which the galleries of these
instructive coal mines are overhung. The roof is
covered as with a canopy of gorgeous tapestry,
enriched with festoons of most graceful foliage,
flung in wild, irregular profusion over every portion
of its surface. The effect is heightened by the
contrast of the coal-black color of the vegetables
with the light ground-work of the rock to which
they are attached. The spectator feels himself
transported, as if by enchantment, into the forests of

another world; he beholds trees of forms and characters now unknown upon the surface of the earth, presented to his senses almost in the beauty and vigor of their primeval life; their scaly stems and bending branches, with their delicate apparatus of foliage, are all spread forth before him, little impaired by the lapse of countless ages, and bearing faithful records of extinct systems of vegetation which began and terminated in times of which these relics are the infallible historians."

But we are told that the Naturalist loses all the poetry of Nature. There is no greater mistake than this. He has become so accustomed to the beauties of nature that he is not ready, like the novice, to utter exclamations of surprise. Bring into a fine gallery of paintings or of statuary one entirely unaccustomed to such works, and he is constantly manifesting his surprise at the novelties, and is perhaps equally delighted with the coarse daub and the work of the greatest master. Think you he enjoys more than the artist, who stands silently drinking in the beauties for the hundredth time from the fine touches which the other never perceives?

Burke was undoubtedly a great man, but he made a great mistake when he said that " our ignorance of nature is the cause of all our admiration." If he had said that our ignorance is the cause of all our *exclamations*, it would have been near the truth. Give to the naturalist his microscope, and let him see new beauties in the wing of an insect or the veins of a leaf that he never saw before, and you hear him exclaiming as others do when they look upon beauties that he has seen hundreds of times. Our exclamations are not signs that we see or appreciate the beauties of nature more than others, but simply that we see them now for the first time.

Wander, then, through beautiful cabinets, and each day they will become more beautiful; go out into the fields and study with care every object there, and you will be astonished at the beauties which God has scattered with such a liberal hand, that scarcely a place can be found where some have not fallen, though unperceived by the hurrying multitude.

LECTURE III.

Natural History as related to Wealth.

It is sometimes pleasant to journey alone, and sometimes we choose the highway where we are sure of companions. In our speculations, we may like to strike out new paths, or at least to travel in those that are unbeaten; but if we would find ready listeners, we must select those subjects on which all in the main agree, and consider those relations which all can readily understand. If we can open a road to wealth, we are sure that it will never be deserted. The riches may be in the gold-dust scattered in the sands of some far-off plain, or in the whale and seal among the icebergs of the northern seas, or in the deep-caverned mines of coal,—the way will be crowded. Hundreds may fail, but others rush to take their places, as though this were the great battle of life, and the watchword were "victory or death." No nation in the world is more ardent in the struggle for these prizes of

money value than ours, if we judge by the eagerness
with which we devise plans, and our willingness to
endure labors. The money value is the one we
oftenest quote, and when we remember its power
we can hardly wonder that we do so. It is a necessary means for the growth of the fine arts, as well as
the moving power of the useful. It renders possible those gigantic schemes by which progress is
hastened, seas covered with commerce, mountains
pierced with tunnels, states joined by roads of
iron, and nations joined with telegraphic cables.
We are not only taking a ground of common interest when we consider the bearing of Natural History upon *Wealth*, but one deservedly so, for it is
important. Should we define wealth as some seem
to do, with good reason, as any thing *which can be
enjoyed or purchase enjoyment*, we should give it a
much wider signification than is usually connected with it; but our work for the present will be
more simple if we give it the common meaning,
which is *money*, or something which money rep-
resents.

The most obvious benefit of Natural History is

the development of new resources of wealth. And
in this respect Geology and Mineralogy stand pre-
eminent. So far as the mineral resources of the
earth are concerned, they can not be over-estimated,
and they are most readily perceived and most eagerly sought for. A portion of the metals and other
valuable minerals are so accessible that they have
been reached by men in all ages. But the amount
thus accidentally found would fall far short of the
present wants of man, and those wants will rapidly
increase every year. The most valuable often appear under forms that would only be recognized by
adepts in mineralogy. Others can be discovered
and followed only by an intimate knowledge of the
structure of the earth's crust. They must be sought
for by the light of science. There is a natural connection between certain rocks and valuable deposits,
as the salt-beds and brine-springs with the New
Red Sandstone. A knowledge of these connections
gives certainty in investigations, and this knowledge
is the fruit of geological study. And should a
mine by accident be discovered, it is only the principles of this department of Natural History that

can determine its value, and give security in the investment of capital.

Any other mineral sinks into comparative insignificance if valued with coal. The whole history of this valuable substance is an argument in favor of the study of Natural History. In searching for it, millions of dollars have been thrown away in boring rocks in which a pound of it could never be found. In fact, the substance which has misled the majority, is itself a positive proof that no coal is to be expected.

By the curious, and I might say wonderful revelations of Geology, vast beds of this substance have been discovered where no accidental discovery could ever be made. And when discovered, their productiveness and method of working are determined by the principles established by this same department of science. This is of yearly occurrence. No country in the world is richer than ours in this and nearly all other valuable minerals. These constitute no small part of our national wealth. And they are destined yearly to become more important because of the increasing demand, the discovery of new deposits, and more efficient and economical methods of working them.   Our coal, our iron, lead, and gold are inexhaustible, and must give us immense resources when fully developed.

Look at England, and inquire the sources of her wealth and power.   Lock up in her hills and valleys her coal and iron and all her other mineral wealth, and you have taken from her one great element of her power.   It is the mineral wealth in that little island, developed by the science of her distinguished men, that enables her to manufacture almost for the world.   Her coal moves the thousand looms and ponderous hammers that load her ships with fabrics and swell her revenues.   It has been referred to as a striking illustration of the influence of mineral wealth, that fourteen of her large towns, from Exeter to Carlisle, are built along the strike of the New Red Sandstone.   To that formation belong the brine-springs and beds of gypsum, and immediately beneath is found the coal.

Her scientific men have scanned her soils and cliffs beneath them, and in them they have found the means of civilization and comfort at home, and the means of commanding the obedience of some nations, the money and respect of all. We have not been entirely wanting in the work of developing this field of wealth. Enough has been done to show that the United States contains some of the richest mineral districts in the world. We may not abound in gems; but where in the world do such beds of coal and mountains of iron abound? We see in them the elements of power; but they must be developed, and the field enlarged by new examinations and discoveries. Our general government has not neglected this portion of its possessions. It has sent out its geological surveyors to examine and locate mineral lands; and they have rendered important service in developing the wealth of the country, as well as in making valuable contributions to science. Our States have also understood the value of these investigations. Large sums have been appropriated by many of them. And in none, so far as I know, has the money failed to yield a full return in kind, besides an immense benefit to the general cause of science. In some, the return in a single year has been a hundred-fold. Such explo-

rations do not produce their best results at all times
in the discovery of minerals. It is to this that attention is most directed, and success in this is generally the criterion by which they are judged.
Other objects receive attention. They point out
general characteristics of soil as derived from certain classes of rocks, discover fertilizers, and thus
give important aid to agriculture. They point out
proper building materials by the discovery of quarries, and clays, cements and paints. Such examinations in many cases give important hints in en-
gineering, the draining of land, the sinking of Ar-
tesian wells, and consequently bear upon the health-
fulness and habitableness of large tracts of land.
In all these incidental methods, Natural History
bestows wealth, without receiving credit from those
benefited.

States have generally shown their wisdom when
ordering geological surveys, by connecting with
them surveys in every other department of Natural
History. The plants, the birds, the fishes, the
quadrupeds, and insects have each been deemed
worthy of study. Many have sneered at the idea of

voting money for "bugs and hornpouts." Very
many have favored such schemes, hoping that a coal
mine, at least, would be discovered on their own
farms; while the birds and fishes were added and
carried along by some shrewd managers, as politi-
cians would say, "like a passenger under the boot."
These departments do not attract attention so
readily, because their connection with wealth is not
so direct and obvious as the discovery and working
of minerals. They are some of them of equal importance, and are destined yet to become of the very
highest value in an economic point of view. These
investigations add to the number of the useful
plants, teach us to protect them, and to increase
their value.

The earth produces more than a hundred thousand species of plants; they are directly or indirectly
serviceable to man. This is true, at least of the
greater portion of them, without doubt. The lovers
of Botany, from the days of Solomon till now, have
been bringing out the beautiful and the useful in
this kingdom of nature. What multitudes of plants
now minister to health and luxury, of which we

should have known nothing but for the special study
of this science of Botany! There have been brought
out those general laws of classification by which
the general properties of plants are inferred from
their structure, so that we can, as it were, read the
labels at a glance which nature has affixed to them,
inviting us to enjoy or warning us to beware. The
history of this science shows that much of the labor
bestowed upon it was simply to classify and name.
System after system—if some of the earlier attempts
could be called systems—has been thrown aside;
and to a casual observer the labor seems to have
been lost—at least so far as wealth is concerned.
This is but a superficial view. Those old pioneers
were groping toward the true goal—their progress
helped on those who came after, their mistakes
warned them; they added one after another to the
list of useful plants. The work commenced by
them has kept its steady course, till now we can
hope for but little more in classification, except on
some isolated points, or in the minutiæ of the details. It remains to perfect local floras—to name
those plants that may be discovered—to turn the at-

tention more generally to vegetable **physiology, to
the** unfolding of their uses. This will follow necessarily when the preliminary work is done of collect-
ing and naming. This has been fast progressing by
the study and labors of those old Botanists, whom
we are apt to think of as remarkable for their zeal
alone.

Enough of each kingdom is accessible to men **for
the supply of their wants in a** primitive, simple state
of society. Those portions they have had longer
than we can tell. But when science commences, its
first work is to classify and give names. When **this**
is done, **the nature of the things is more carefully
studied for new principles of classification, or to**
confirm the old. This very process brings out the
useful properties of some, and the noxious character of others. And when this work is completed,
the possible *new* **uses are the regular, almost the**
necessary, subject **of study. To that point we are**
fast coming in **the study of the vegetable** kingdom.

The **triumphs** of this study thus far, and its pro-
phetic achievements, are graphically given by the
poet :

"There be flowers making glad the desert, and roots
    fattening the soil, ...

And uses above and around which man hath not yet
    regarded.

Not long to chase away disease hath the crocus yielded
    up its bulb,

Nor the willow lent its leaf, nor the night-shade its
    vanquish'd poison;

Not long hath the twisted leaf, the fragrant gift of
    China,

Nor that nutritious root, the boon of far Peru—

Nor the many-colored dahlia, nor the gorgeous flaunting
    cactus,

Nor the multitude of fruits and flowers ministered to life
    and luxury;—

Even so there be virtues yet unknown in the wasted
    foliage of the elm—

In the sun-dried harebell of the downs, and the hyacinth
    drinking in the meadow—

In the sycamore's winged fruit, and the facet-cut cones of
    the cedar.

And the pansy and bright geranium live not alone for
    beauty;

Nor the waxen flower of the arbute, though it dieth in a
    day;

Nor the sculptured crest of the fir, unseen but by the
    stars—

9*

And the meanest weed of the garden serveth unto many
    uses,—

The salt tamarask and juicy flag, the freckled orchis and
    the daisy.

The world may laugh at famine when forest-trees yield
    bread,

When acorns **give** out fragrant drink, **and the sap of the**
    linden is as fatness ;

For every green herb, from the lotus to the darnel,

Is rich with delicate aids to help incurious man."

To accomplish all we wish and all we expect, in
bringing the vegetable **world** to render its riches
more abundantly, we must undoubtedly call to our
aid the kindred **science of** Chemistry. But here is
the great storehouse of materials. All **our** food
comes directly or indirectly from the vegetable
kingdom. The root, the leaf, the flower, the fruit,
the sap, each in turn in various plants, constitute
directly the great mass of our sustenance. **And the**
exception, when we use animal **food, is** only
apparent, for every animal used for **food,** from the
oyster to the ox, is directly or indirectly dependent
upon plants for his subsistence. All animals, man
included, are so constituted that they can not **subsist**

upon inorganic elements. We may analyze our
food, determine its exact composition, but it will
not enable us to feed on minerals. We may prove,
with all the science of a Liebig, that charcoal and
air and water contain all we need, but we know
they would form poor fare for our tables. We may
call in the aid of Chemistry, with all its power to
produce transformation—give it a magazine of the
pure elements—and it can not furnish us with a
single grain of starch nor crystal of sugar, nor with
any thing to be a substitute for them. The plants
are the only chemists that can take up these
inorganic materials, and in the wonderful laboratory
of their living tissues mold them into forms to
support animal life. All that I have said of
nutritive plants might also be said of those having
medicinal properties, and of use in the arts. Our
fine fabrics, our brilliant dyes, our most grateful
perfumes, come in a large proportion from this
kingdom. Here have been found those wonderful
modern-discovered substances—India-Rubber and
Gutta-Percha. How long these and other valuable
products remained unknown! How many more are

to be discovered, as the wants of men demand them,
and the study of plants is more thoroughly and
and generally pursued! To investigate the laws of
that department of nature upon which all animal
life depends, is certainly an imperative duty. But
more than this, for our present purpose—we see this
kingdom the channel by which many of the luxuries
of life are poured in upon us, and the only source of
many materials necessary to the present state of
civilization. The fiber which clothes us—upon
which we print our books—the gums that surround
our submarine cables and take a thousand forms of
usefulness, are already sources of national as well as
individual wealth. To increase these products and
ward off diseases to which valuable plants are liable,
are the direct results of thorough scientific investigations in the various departments of Botany. All
our knowledge has failed to arrest the blight of the
potato; but all feel that the more perfectly we
understand vegetable physiology, and all the habits
of particular plants, the better we shall be prepared
to improve them in quality, to increase their
quantity, and protect them from injury. Science

has not failed, but we have failed for the want of science. As we bend our minds to patient study and careful observation, we may be able to arrest disease in plants, improve those already useful to man, and discover valuable properties in thousands now apparently useless. Many of our valuable fruits were once entirely useless or noxious. That they have been brought to perfection, or that all those capable of such improvement have already been pressed into the service of man, we have no reason to believe. In fact, the progress made every year, and especially the progress made the last twenty years, gives great promise for the future. This rapid progress has been made because those engaged in Agriculture and Horticulture have worked by the light of science. When we see beautiful nurseries and gardens, we shall find in the owner or keeper the knowledge of the science for which we are here contending. If they have not the broad principles, we shall find them acting by the rules of some broader mind, who is at home with Decandolle, and Lindley, and Lowdon, and Gray. We can hardly overestimate the advantage to our own country, if all our young men who travel, our consuls and missionaries, were so versed in science that they should be able at once to detect the valuable properties of plants and their habits, that all capable of introduction might be secured at once. A single plant might repay for all the time and labor of every American student in this depart-ment. But if men are never trained, they do not observe. And if a strange plant is forced upon their attention, they know so little that they can determine nothing of the prospect of improving its qualities by cultivation, or even of cultivating it all. If all those who labor among plants, and have opportunities of introducing new, were well versed in Botany as it is now understood, this source of wealth would be vastly increased in a single year. The progress would be rapid. The quality would be improved, and the number would be increased. Useful plants would take the place of those useless or noxious. Our forests would be better preserved, and new forests would be springing up on rocky hills and neglected swamps. Millions of acres, bar-ren and dreary, might be gradually supplying our waste of trees, if men would learn that forests can be planted, and were imbued with that spirit of improvement and care for coming generations which science has ever had a tendency to produce.

It may be said, with truth, that much of the work already done has been done by those ignorant of science. These results have been the slow accumulation of ages; we wish now more rapid progress. The times demand it. The same is true of every department of human industry and source of wealth. Discoveries in olden time were accidental. The Alchemists in the dark ages, with their alembics and crucibles and chemicals of mystic names, worked by chance, and by chance, from time to time, made some valuable discoveries. But how different is the work of a modern chemist! A thing is to be done, and he is able at once to bring to bear upon the problem all the principles of that wonderful science. Every experiment is performed for a definite purpose, and accidental discovery is the exception and not the rule. So in mechanics—a result is to be reached, and the problem is attempted by well-established principles. Those wonderful

looms that ply their iron fingers to weave our
carpets were not a chance discovery by Bigelow—
they were an invention, reached only by long-
continued systematic study. So of discoveries and
improvements in the vegetable kingdom in our day.
They must not be left to chance, but be sought for
under the guidance of science, where alone the
course is direct, the progress sure and rapid.

Perhaps Zoölogy does not give promise of so rich
a return as Botany, in material wealth. We do not
expect to discover important animals for domestica-
tion, nor do we expect to add very many valuable
animal products to those now known, by the discovery of new animals. So far as their products are
rendered more useful or increased in number, we
shall probably be indebted to Chemistry, rather than
to the pure science of Zoölogy. But there are important indirect advantages that may result from it.
The study of the structure of the whale renders it
highly probable that there is an open polar sea. It
is, as Professor Agassiz remarks, perfectly convincing to the physiologist; if the whales in winter are
not all south of the frozen belt, they must find open

water beyond at the north. This opinion of the learned Zoölogist will undoubtedly stimulate these explorations until that problem is fully solved. If, then, this surmise proves to be correct, and the whale-fisheries can be carried on successfully in that great northern ice-bound field, it would certainly be a remarkable instance of the indirect benefits of science. We may be asked to wait until the discovery is made. We only refer to it as a thing so conclusive that action may reasonably be based upon it, and the grand result may reasonably be expected to follow. What connection would there seem to be, to one unacquainted with the history and bearings of all science, between the study of a whale and the discovery of a northern sea, and the establishment of productive fisheries? Whether this may be realized or not, such are the constant results of science, from subjects that in the beginning gave promise of nothing but the gratification of curiosity.

The study of the habits of fish will enable us to protect them, by law, from those methods and times of capture which prove destructive to large numbers without any adequate return. It gives us also the prospect of being able to stock our lakes and streams with **valuable fish, as** easily as we can supply our farms with flocks, and to much more profit. **The time is not far distant,** when those who have sneered at the study of "eels and mudpouts," and have made speeches on economy when States have appropriated money for this purpose, may find that there are some things not dreamed of in their philosophy, and that money can be made where they never suspected it.

The study of the beautiful birds and the hideous reptiles has corrected many false notions, and **shown** that the former, at least, are a flying guard for **the** protection of our fields and gardens. We are glad to invite their aid, and divide our delicious fruit with them, that we may save the remaining half from the insects. **Even the crow,** despised **and persecuted** as he is, is found to **pay well for the few** grains of **corn he may steal.** We come by careful study of all these classes of animals to learn their true place—to learn the use we can make of each of them—the methods of protecting the useful and **of**

guarding against the injurious. The money value
that comes from such a knowledge amounts, in our
own country, to millions in a year, and what it might
be to the whole world is beyond computation. And
this knowledge is every year becoming of more importance. We can not, perhaps, select a better example for the perfect illustration of what I mean
than insects. Men, who pride themselves upon their
wisdom and common-sense, and pecuniary shrewd-
ness, generally regard Entomology as a very ridiculous subject; they have never attended to it, indeed it would be the last thing they would think of
doing. To see a man catching bugs and butterflies
is, to them, more senseless than studying frogs and
sticklebacks, if possible. Let us, however, stop a
moment, and inquire what this busy tribe produce
and destroy in a single year. Of their productions,
we may mention the silks, the wax, the honey, the
lac-gums and dyes, the nut-galls and cochineal.
How many millions of dollars, think you, would
purchase all these products for a single year? To
narrow the question, how many millions would buy
those imported into this country, and those produced

here, for a single year? We should hardly be willing to give up our portion of these products. There are many others that we could better spare, so far as comfort and ornament are concerned. They constitute an important item among the necessaries and luxuries of life. Now it can hardly be doubted, that study of this department of nature would tend to increase the quantity of these products, and in some cases improve their quality, and that others of importance may be discovered. This view, alone, would certainly remove Entomology from the rank of useless and merely curious studies, to one having important bearing upon comfort and health.

But insects are also destroyers, and this to an alarming degree, and their ravages in our country are yearly increasing. It was some time since found that their injury to the crops, in this country, amounted to more than twenty millions of dollars a year. I think the same report made the remark, that, if a foreign nation should injure us the twentieth part of this sum, for a single year, our army and navy would be called into requisition to de-

mand and obtain satisfaction. We should all sustain the action—but we sneer at bird-laws, allow
insectivorous birds to be destroyed for sport, and
regard those who study insects as foolishly em-
ployed. It is by the labors of Harris, and such ob-
servers of this hungry, numerous host, that we can
drag them from their lurking-places, know them
under all their disguises, destroy the injurious by
taking advantage of their own instincts, and spare
those that are useful by preying upon others. Birds
are our natural protectors from this foe. But the
broad acres of cultivation have increased faster than
the birds. Our only help is science—the study of
their whole Natural History. This will save for us
millions in a single year. In some parts of our
country the struggle is now really a desperate one—
many choice products are preserved only by con-
tinual warfare. And to maintain our ground, we
need the aid of every entomologist in the land.
The labors of such a man as Harris are worth many
fortunes every year.

We are all ready to acknowledge that Agriculture is the grand source of national wealth. It is

an evil day for any country, when this calling falls
into disrepute, or is neglected for other more alluring
and perhaps quicker sources of wealth. We have
already indicated how the study of Natural History
lends important aid to this branch of industry, by
introducing new plants, and giving more perfect
knowledge of their habits, the methods of improving their quality, and of protecting them from
injury. But were this all, it could not give it that
dignity and success which we believe it now con-
fers. It is fashionable to laud farming, but facts
seem to indicate that for some years past it has been
unfashionable to engage in it, where men must
labor with their own hands. It is not the labor that
has driven them from the field, for they have left it
oftentimes for more laborious and exhausting pursuits. Go through that large portion of our country
where those who live by cultivating the soil must
labor with their own hands, and inquire in every
family what business they intend for their sons, and
you will find farming to be the exception and not
the rule. One is intended for some trade, another
for the counting-room—another for Law or Medi-

cine—another is sent to college, trusting to chance
to direct to some subsequent employment. And
should he choose to be a farmer, his parents and
neighbors would most likely consider his college
education as thrown away. By our words, then, we
praise Agriculture, and by our practice we condemn
it—brand it. And for both we think a satisfactory
answer can be given. Agriculture ought to be the
high, noble, and honorable employment which it is
represented as being in our agricultural addresses.
It deserves to be, and might be; but then the
question at once arises—if it is, why is it that almost
all men, even farmers themselves, are so anxious to
secure other business for their sons? We are
constantly affirming that Farming is as honorable as
Law or Medicine, and yet it seems hard to make
the world believe what they are constantly assert-
ing; for there are but few farmers who would not
rather see their sons eminent doctors and lawyers
than good farmers. This ought not so to be—for
tilling the soil is undoubtedly a natural occupation,
and therefore ought to be made desirable. It is
well for us to look for the evil, and correct it. The

low estimate of Agriculture is undoubtedly due to
this fact, that less thought and study have thus far
been needed in this than in most other pursuits.
The prospect of making money will alone induce
many to engage in certain pursuits for a time. But
look over the pursuits which men engage in for life,
and you will find, as a general rule, that the *thought*
required to carry on a pursuit is the measure of its
dignity, and the index of the class of persons who
will engage in it. Men of learning and thought
and refinement, can never be induced to engage in
any work that can as well be carried on by men
"who never had a dozen thoughts in all their lives."
If a railroad is to be built, the engineers will labor
hard to make the surveys and measure the grade,
because that requires thought; but they will not
shovel sand for the same price, because that can be
done as well by the unlettered Irishman.

All labor becomes honorable and dignified just
in proportion to the intellect, the thought, and study
required to carry it on. These render base things
noble. "The chemist's and geologist's soiled hands
are signs of no base work; the coarsest work of the

laboratory, the breaking of stones with a hammer, cease to be mechanical or ignoble, because intellectual thought and principle govern the mind and guide the hands." According to this principle, to which, we believe, all will assent, every source of study and thought which we can connect with agriculture will give it dignity and attractiveness. And just in proportion as these are wanting will men relinquish it, if possible, as they become intellectual and refined. We have only to make agriculture require as much thought as the learned professions, and men will need no panegyrics from agricultural orators to induce them to forsake the counting-room and office for life in the open air. Nothing can produce this desired result like the study of Natural History. Perhaps we ought to add Chemistry. But this requires such skillful manipulations that it must be confined mainly to a few who make it a profession. But not so with Natural History. Every portion of it can be made practical and of interest. *Agriculture is Natural History applied.* Geology, Botany, and Zoölogy are its basis, and in proportion as these are understood, will there be success. It is because these sciences are the basis of Agriculture that men have theoretically considered it noble; it is because it has to a great extent *ignored* these sciences, its true basis, and become a changeless routine, that it has practically been considered base. When the farmer studies the minerals of which his soil is composed, the plants that spring up around him, the insects that destroy—when he learns to study all the objects which abound on every hill-side and valley —farming will be a science that will daily awaken thought, a pursuit in which mind can develop, and then it will not only be among the most honorable, but the most honored of secular professions. Just in proportion as it takes this place does it rise in dignity, and call men of culture from other pursuits to this.

So far, then, as we look to the improvement of agriculture in all its departments as a source of wealth,—and all acknowledge it to be the most important—in fact, the only sure basis,—just so far do we acknowledge the relations of Natural History to wealth, and make apparent the need of study in

every department of this division of science. All
men will encourage those departments which will
bring money at once.   But we see a very different thing is needed : it is to make every plant,
and bird, and insect, every object of Natural
History, a subject of thought—that the field may be
a place of intellectual as well as of bodily activity.
This may be thought impossible, but it is not.   We
here see the need of certain kinds of information
which some undervalue.   When our agricultural
reports give the Natural History of an insect, the
picture of a bird, or a snake, a grass, or a sedge, it
is often a better work than reports on wheat or
stock, however valuable they may be.   These
objects, from the forest and the river, turn the
thoughts into a new channel, and waken powers
of observation, that, but for them, might ever have
remained dormant.   Much work of this kind done
by our national and state governments, that has
been hastily, though undoubtedly honestly con-
demned, has its value.   What use of describing
fossil shells, in boundary surveys? grasses and birds
in astronomical expeditions, or corals in the coast

survey? many are ready to ask—as though it were a waste of money, or at least a poor return for it. But go through our country, and see such books studied by thousands of the young, who but for them would have never had a thought awakened respecting such objects, and we shall be satisfied that they are no waste—no mere gratification of scientific men—but the educators of thousands, and will, in the end, not only elevate, but return far more than their money value.

In this view of the subject which we have presented, that thought dignifies labor, we see why farming was more honorable among the ancients than among the moderns. They honored it practically, while we profess to do so. We think the reason is at once apparent, and illustrative of our position, when we compare farming with the other pursuits of those times. It came nearer to the learned professions than it now does. When we consider the state of the other sciences, and see also the knowledge of Agriculture displayed in the works of Virgil and Cato, we find it to be *the science* of those times. It was not pursued by the learned and brave of those days merely as a matter of profit, but because they found in it the best sphere of refined and intellectual pleasures. Thus, when Cicero introduces Cato in his *De Senectute*, he causes him to say that he cultivates the earth not only as a matter of duty, because it is beneficial to the whole human race, but because it is a source of delight. He is delighted with that secret power which, from the minute seed brings up the tall trunk and wide-spread branches. The preparation of soils, the pruning and grafting, are to him sources of pleasure, and are deemed honorable employment for kings themselves. We need only raise the cultivation of the soil to its former comparative position with the learned professions, to make it as highly esteemed. This can never be done by praises at every agricultural fair, but it can be done by encouraging every department of Natural History, until they shall make it as fine a field for intellectual enjoyment as either of the learned professions. What is now a mere drudgery might become a delightful employment, and the labor receive a more certain and abundant reward.

11

We increase wealth, when we change to means of
enjoyment what had before been useless.   Our hills
may be filled with riches, but if we can not recognize the precious ore, we are as poor as though it
were pebble-stones.   The blind man is unmoved by
the beauty of his landscapes—they might as well be
rough and dreary as beautiful.   The deaf man is
none the happier for sweet sounds, that others
would pay lavishly to hear.   A cultivated taste dis-
covers sources of enjoyment where the unrefined
would be like the blind man among pictures, and
the deaf among music.   The connection of Taste
with Natural History we have already discussed.
It is able to throw around even the poor man more
means of enjoyment than wealth can purchase for
the uneducated and unrefined.   If every young
man would acquire such a taste for these studies,
such a love for the beautiful that, by his labor or
direction, a single acre of ground should be rendered more productive or attractive, what an advantage to himself and the world!   It is a great ac-
quirement to be able even to rightly appreciate and
enjoy the labors of others.   The lover of Nature's

beauties owns property in every landscape that his eye rests upon—oftentimes more than he who holds the title-deeds.

Give to the true lover of Nature a hard, rugged soil, and he will know how to make even that attractive. He studies every object. He knows upon what the pleasing effect depends. It may be a single tree, or a single copse, which the thoughtless world never think of sparing. His home may be poor, but it will have the best location, and it will have expression—not a mere roofed box for shedding rain, without proportion, without the first element of beauty, so devoid of taste that every ornament makes it more unpleasing. To most of us, our homes are our wealth. We seek for money that we may throw around us objects of beauty and make our homes places of enjoyment worthy of a civilized and cultivated people. Nothing is plainer than that some people, almost without money, succeed in this, while others, whose checks are readily honored by their bankers, entirely fail. They have the money, but are entirely unable to purchase the same means of enjoyment that the poor man has always within his reach, who is able to select the beautiful objects which nature presents, even in the least-favored locality. Look at two homes of men of equal means. One tasteful in form and beautiful in location—its shaded walks, and every useful tree, and shrub, and plant, arranged with regard to beauty; every natural defect of landscape is softened, every beauty heightened. The other is placed by chance, without symmetry or proportion; no plant of beauty is spared, and in all the surroundings not a single thought displayed that any natural object has beauty. The owner may be conscious that his neighbor has a great advantage, but thinks it his fortune, and no more thinks that he can secure the same for himself, than that he could dig from the earth hidden coffers of wealth. He knows there is beauty, he hears it repeated by every passer-by; but he can not put forth his hand and secure it for himself, though the material surround him on every side, simply because his mind has never been trained to perceive the beauty of separate natural objects. He is aware of it only when they are grouped by others; and then, perhaps, he despises them, or at least undervalues them, because he is as blind to their beauty as the eye that never saw light. Who can compare the worth of these homes as places of rational enjoyment, or the capacities of their owners to enjoy? The man of taste may not have struck a blow harder nor more frequent than his neighbor, but he has had unnumbered sources of enjoyment the other had no power to avail himself of; and now the tasteful home finds ready buyers at liberal prices, while on the other land the buildings are considered rather as an incumbrance, and the soil as stripped even of the materials of improvement which Nature almost everywhere scatters on land untouched by man. The worth of our homes must depend mainly upon the beautiful objects of nature that we can throw around them—at least this must be true of those whose wealth is not abundant. These objects can only be selected and appreciated by that training of the senses, and those ideas of the beautiful, which Natural History studies alone can fully secure.

11*

LECTURE IV.

NATURAL HISTORY AS RELATED TO RELIGION.

WHEN the oak spreads its sturdy branches, and strikes its roots deeper among the cliffs of the mountain, there is one work to which all its changes are preparatory, and this is the production of fruit. In the whole vegetable kingdom, with all its varied beauty, every force and every change is subservient to this higher work. The architect also lays his foundation, but it is only that he may build upon it. So God has broken up the crust of this globe, and covered it with a succession of living forms, but it was only that he might thus the better fit it as a dwelling-place for rational man. To man He has given an intellectual and an emotional nature, but it is only as a condition for that higher religious nature, in which man approaches most nearly the perfect image of God. All nature is indeed made to minister to the physical enjoyment of man, but the wonderful plan of its frame-work is a fit counterpart of his intellectual nature.   Its beauty of adorning, its grandeur and sublimity, are the visible heaven of his emotional nature.   But its highest adaptation, and that to which all others are subservient, is to man as a religious being.   So complete is it in this respect—so fully is a God, and a God for worship, shown in even the humblest plant that clothes the earth—that, to the sincere inquirer,

"This world....becomes a temple,
And life itself one continued act of adoration."

On the common argument from special adaptation for the existence of a God it is not necessary to enlarge, because it has been so fully presented by the ablest writers that it is probably familiar to all. So far as we attempt it, of course we shall confine ourselves to Natural History illustrations; and we do not shorten the argument from any want of materials, for they are abundant—sufficient to present that argument in its full force.   Some have considered this proof from natural objects unsatisfactory; but on this point we may say, that however philosophers may speculate, there is in Natural

History a general harmony, such as ancient philosophers saw in other departments of nature, and regarded as proof of an intelligent author; and to the common mind the argument from special adaptation
will always be convincing—far more so than those
higher speculations and proofs from our mental
constitution. So long as men can observe Nature
more easily than they can study their own minds,
so long will they be more convinced by the general
argument, as presented by Paley, than by the
intellectual and moral nature of man, which some
consider the only proof of a personal God. There
is, to say the least, a charm about the argument, and
it seems to us to have force. When we see special
adaptations, not occurring once merely, nor in one
kingdom, but in hundreds of instances—adaptations
that we might never have thought of, but acknowledge to be worthy of the greatest genius—the
mind goes up to a personal God. The mental
philosopher may stand there and utter his warning,
he may say that the whole argument is a begging
of the question; the answer practically will be this
—"There are in nature adaptations worthy of the

highest powers that we attribute to a person—their
number is so great as to preclude the idea of chance
—there is, therefore, a person—a personal God."
Take from the animal kingdom a single illustration.
The honey-bee, the wasp, and the hornet, build
geometrical six-sided cells. This form is best fitted
for their purpose—they necessarily build them in
this way, either compelled to it by their organization or instinct; for our argument we care not
which. These three tribes go on building the same
way forever. Now, how came that geometrical
form to be selected for those three insects? By
whom was it done? Only one answer can be given
—by a being capable of considering all possible
geometrical forms in the abstract, and of selecting
from all possible forms one best fitted for these
three tribes of insects. No one will pretend that
these insects were from the same stock, and thus
account for the common form of their cells. The
materials are varied. By the bee the form is
produced in wax secreted beneath the rings of the
body, by the others in paper, formed of the woody
fiber of our fence-posts and door-sills; but when we

direct our attention to the *form*, we see **evidence of
the** highest intellectual power in considering **ab-**
stract geometrical relations. Such arguments can
be **repeated** almost without limit, and if there is
failure to convince, it seems to arise from the defect
in the method of studying **the proof,** rather than
from its nature. If asked, then, why this argument
from adaptation has not been more convincing, we
answer, it has not generally been studied in the
right way. It has been studied in *books* rather than
in the *field*. The effect of this **we** shall notice
farther on.

We are inclined to reverse the order of the argu-
ment presented by Paley, and give the vegetable
kingdom precedence—make it the strongest link in
the grand chain of proof. He remarks that " a
designed and studied mechanism is in general more
evident in animals than in plants, and it **is unneces-**
sary to dwell upon a weaker **argument when a**
stronger is at hand." There are many points of the
whole subject that have changed their relative
importance, in fact their whole bearing, since his
day. The unity of plan in the whole vegetable and

animal kingdom, so fully brought out since his time, has to many a stronger bearing than all the special adaptations so clearly presented by him. His first sentence, "That if in crossing a heath (*suppose*) he pitched his foot against a stone, and were asked how it came to be there, he might possibly answer that, for any thing he knew to the contrary, it had lain there forever," is a landmark in science, and shows the wonderful changes since his time. That same pebble would bring to the mind now the vast forces that have spent their fury on the crust of our earth—all that war of the elements by which the earth was fitted for our dwelling-place—all the movements of that vast plan which has been moving on for countless ages, when no eye but that of God looked upon the scene; of which no record is left, but that engraven upon the hills, and scattered with the pebbles,—in Paley's day an unknown language, now rendered vocal by the light of science, as the statue of Memnon gave forth sweet sounds of music when lighted by the rays of the rising sun.

In the vegetable kingdom we see special adaptations, prospective contrivances, and yet there is no thought, no instinct even, to guide. The dreaming philosopher might talk of feet becoming webbed by attempts to swim—of wonderful changes produced in ages by the law of progressive development. The followers of Oken and Lamark, without the science of their masters, may believe that their ancestors were fish, and that they are not themselves denizens of the deep because some enterprising member of the family floundered out of the water and forgot to return; but even this accomodating theory will give no explanation of all the wonderful adaptations of parts by which individual life is carried on, and the species propagated, in the vegetable kingdom. The monad, by its desires, may be fancied to pass through the varying stages of oyster, fish, and ape, up to man himself; but that the one-celled plant of our northern snow, or those that abound in our pools, should suddenly become ambitious, and be satisfied only by spreading like the oak or blooming like the roses, is a far more difficult problem for their accomodating philosophy.

Did you ever look into a single flower, the lily

or the rose, to see there the wonderful machinery fitted for a specific work, the production of the seed? In the center of the flower, completely surrounded, secured from danger, are the first sketches of seeds—now mere points, but each one fitted to receive an independent life. It has not yet come, but the home is prepared for its reception. And now another portion of the flower, the trembling stamens, that seem tipped with golden points to draw down the spark of life from heaven, give out the gathered force locked in the floating, dust-like grains of pollen. They strike the central organ, and, as though drawn by an invisible power, thread their way down long tubes and touch each seed with the fire of independent life. The seed, no longer a mere cell without life, now asserts its dignity, and the parent plant, as though conscious of its precious treasures, gathers with every power the materials needed for the future growth of the young, now cradled in the flower. And around that germ of life the tree collects of its richest products the salts, the starch, and the sugar which form the bulk of the fully developed seed, making it a store-house

12

of food sufficient for that germ when **thrown from**
the parent stock, till it shall put forth roots **and**
leaves, and be able to compel the earth and air to
minister to its wants. And when the acorn drops,
or the grape-seed matures, **what can** you see, with
the aid of your keenest **scalpel and most** perfect
glasses, that shall show you that the work is **com-**
pleted, without a single mistake, in all the countless
myriads that fall in every valley and on every
mountain-side? But in one is a force lodged that
shall send up the stout trunk, spread its branches,
expand its leaves, and produce its fruit, a perfect
oak, and from the other shall come up the leaning
stem to climb the oak with loving tendrils, spread
its thick foliage among its branches, and mingle its
rich clusters of purple with the humble russet of the
acorn in its cup.

Or **go** with me to the field, and watch the setting
of the golden rows of corn. From **the** shaking tas-
sel falls in every breeze a shower of vital dust, and
from the center of the husky ear each half-formed
kernel throws out its line of silk to catch the floating cells of life. And as it gathers in its portion

the grain begins to swell, to gather richness from the parent stalk, till it gleams in southern fields with the softness of the pearl, and in the north with the yellow of the topaz and gold. No parent, with the wisdom of man, can more perfectly provide for its young, than the trees of our forests and the grasses of our fields for the young plant in every seed they mature. If by chance the grain of pollen fails to reach the seed, no germ of life is there, no food is needed and none is garnered up; the tree never mistakes and collects food where its own young is not present to feed upon it.

And when the seed is formed there is still another care, that it may find its proper place of growth; the means are fitted to the need of the plant. To one seed are given wings that it may fly away, the crane's-bill scatters its seed with a curious spring, the thistle rises on its fringed balloon, and others cling to every passer-by and thus are scattered over the earth. And when that seed has germinated, every leaf has a thousand mouths to drink in the gases from the air, a thousand points below the surface of the earth to gather materials there. Every breeze that moves its leaves feeds it, the rocks crumble beneath to give it strength. And as it rises, every change shows its adaptations to all the forces that surround it.

Go through our northern forests and look at the broad-leaved trees—the maple, oak, and elm. In summer they are filled with foliage, on some of the largest are acres of foliage. Now look at their spreading and dividing limbs. Did they hold their leaves, a single winter's snow would split their branches from the trunks, destroy their beauty, and in the end they must perish. But the first frost of autumn paints the green leaves with gorgeous colors, and the autumn winds shake them from the trees, that their naked limbs may be presented to the frosts and ice and winds of winter. In summer, they must have the broad leaves to drink in gases from the air; in their winter's rest they would prove their destruction, and they shake them off, and not a single broad-leaved tree, in our northern climes, holds its foliage in the winter months. Look now at the evergreens, the spruce, and fir, and pine, with needle-leaves, and with trunks that are single shafts that never divide. Every limb is small, and driven, like a pin, toward the center of the tree, so that should it break no harm is done to the general structure. These keep their leaves, and enliven our winter forests by their green. The deciduous trees, like mariners fearful of their strength, furl their sails at the first rising of the tempest, while the spruce and the fir, as though conscious of their strength, spread every stitch of canvas and bid defiance to the storm.

Every tree, we believe, in its special adaptation, shows a personal God. A single seed of the dandelion, floating on its delicate balloon, would seem to be enough to cut up all atheism by the roots. To some, these proofs may not be satisfactory; but are those who can not see the proof, sure that they have seen all the beauties and adaptations that every day open to the active naturalist? Is it not possible that there should exist in Nature some proof of a creative mind, besides the mind of man? No common, casual view of Nature can justify a negative answer. Things never seen can not convince. It is easy for one who has never seen a fossil, to believe men mistaken when they **talk of** splitting fishes from solid rock—or to doubt **that** coal is of vegetable origin, if he has never visited a coal mine. But when he **walks** among the rocks his skepticism vanishes. **And, on the** other hand, things always seen cease to have their proper effect. If we admire the striking objects of a foreign **land,** we shall find those who dwell among them as un-**moved as** we are by the common objects of our **daily life.**

The effect that common things might have, if presented for the first time, is beautifully illustrated in **the fragment** of Aristotle preserved by **Cicero** in his *De Natura Deorum.* "If," said he, " there were beings who lived in the depths of the earth, in dwellings adorned with statues, and paintings, **and** every thing which is possessed in rich abun-dance by those whom **we esteem fortunate; and if** these beings could receive tidings of the **power and** might of the gods, and could then emerge from their hidden dwellings through the open fissures of the earth to the places which we inhabit—if they could suddenly behold the earth, and the sea, and the

vault of heaven, could recognize the expanse of the
cloudy firmanent and the might of the winds of
heaven, and admire the sun in its majesty, beauty,
and radiant effulgence; and, lastly, when night
vailed the earth in darkness they could behold the
starry heavens, the changing moon, and the stars
rising and setting in the unvarying course ordained
from eternity—they would surely exclaim, there are
gods, and such great things must be the work of
their hands."

These wonderful works have been ever before us,
so that it is hard to realize that there was a time
when they were not—and harder still to feel the
full force of the proof which their mechanism ought
to be to us. And the humbler objects of Natural
History, not calculated to excite emotions of grandeur and sublimity, which we daily tread beneath
our feet, would, according to the common laws of
mind, pass unnoticed, or when noticed fail to convince us as they ought. There may be a wonderful arrangement of parts, all fitted to produce a
certain result; but then we can not see the hand of
God tinting the flower, and arranging each part for

its appropriate work. The plant springs from the ground, and its kind has done so for thousands of generations. If we could but for a moment see the Divine hand apply the rule, weigh the elements, and join the varied cells, how changed the argument would be! But from the *work* the builder must be known. As we walk among old ruins, it is hard to realize that the stones were hewn and raised and joined by men. When the American first visits Mount Vernon, how difficult to realize that here really is the home of the great hero whose name he has ever revered.

It is not strange, then, that this difficulty of realizing should, in the case of natural objects, sometimes end in doubt of a personal God. It is not strange, at least, that it should be so to those who see no more than they saw when children—the merest fragments of the common forms that surround them. And though the wondrous works of design should be described, it is not he who studies them in books, but he whose eye has seen the living loop and hinge that can understand their power to convince. What knows the man who has merely read of Mt. Washington, of the sense of power he feels who climbs the titan blocks which form that grand monument of Nature's forces? What knows the man who has simply read of Niagara, of the emotions of him who looks up to the bending flood, and is deafened by its thunder? It is the real thing, and not its description, that must be relied on to convince. And if we wish to prove the strength of the argument from design, must we look to those who have only read and looked upon the same unvarying surface all their lives, or to the naturalist, who has been walking within the temple of Nature all his life, each day opening some alcove filled with new beauties and adaptations? Shall we inquire respecting the landscape in the distance, of him who has always walked upon the plain at the base of the mountain, or of him who daily ascends that mountain, and views that landscape from every possible point? The common observer is like Aristotle's fancied beings in the center of the earth —remaining there forever, hearing of the gods and their works, but seeing the whole array of Nature only as delineated in pictures of landscapes, and the

orreries invented by men to represent the movements of the heavenly bodies.

But the naturalist, with his trained senses for observing, is, as it were, raised from the center to the
surface, to look off upon a new world. And with
his microscope a new world bursts upon his view
every hour, not as a far-off star, threading its way
like a point of light through the heavens, so that its
motions alone can be determined—but in the drop
of water, in the grain of slate, in the scale of fish,
and every fragment of bone. Think you that
argument from design had no force with Cuvier,
whose firm belief in final causes led him to such
splendid results? His genius called back the perfect forms from the bony fragments in the Paris
basin; but the firm belief in special adaptation
was the guiding light that led him on, and he
never once doubted that the plan and form of each
organic being were fixed by an intelligent, divine
Lawgiver.

Think you the force of that argument is not felt
by Agassiz, as a single scale reveals to him the
character of the fish? We need not debate the

question, but appeal to him, and his answer is given out in his last great work, the ablest volume ever written in proof of the being of a personal God from Nature—the arguments all drawn from the animal kingdom. Here, then, is the man whose eye has seen more forms in Natural History than any other that ever lived—one of the greatest naturalists in any age, the stamp of whose foot can almost call up the forms from their sleep of ages in their rocky beds, seeing in all the adaptations overwhelming proof of *mind.* Here, then, we may rest the argument for design, disregarding the attacks which dreaming development theorists may make. This argument will need no farther defence till it is attacked by some naturalist of such vast acquirements as to make him a foeman worthy of Agassiz's steel.

We are perfectly satisfied with the argument from the mental constitution of man. But if the inside of a watch prove design, does not the case, and all the outer works, fitted to protect the inner mechanism, and reveal its movements? And if the mind of man show design, does not the body, the home of that mind, and the thousand contrivances in nature for keeping that body with all its complicated machinery in tune? If, then, we grant personality to the mind of man, we may, so far as the argument is concerned, believe with some of the old philosophers, that mind to be eternal, uncreated ; but the fitting of a body to the wants of that mind, would prove personality on the part of the creator of the body. There are those who believe that the Saviour of the world was not man in any true sense, that the divine, eternal, uncreated Mind was united to a human body. There is no *absurdity* in this view. And if every man were considered the same, the body, in its adaptations to the mind, would still require a creator equal in kind to the mental part provided for.

But among the rocks of the earth has Natural History laid a foundation for Natural Religion, one that can never be weakened, but is becoming more firm by each new discovery. This it does by carrying us back to the beginning of all organic life, and by pointing out, on the rocky chart, where each new form commenced its course. The infidel argument of an infinite series, might be combated by metaphysical argument, but the reply was only an argument of words, and to many minds far from being conclusive. It was a rampart, behind which thousands would vaunt themselves to be safe, and, like all metaphysical ramparts, so long as it was firmly believed in, it was safe.

But geology has a shorter and more conclusive answer. One blow of her hammer and the rampart of infinite series dissolves as by enchantment. She points her finger back to the granite frame-work of the globe, and reads a chapter in its history, when organic forms were impossible on the molten, glowing mass. Then through each of the stony layers, she marks the introduction of each new species— the thousand wonderful forms, each a miracle of creation. First she unfolds the varied forms of the Silurian seas, the earliest types of organic existence, the chambered shells, the mountain masses of curious patterns, whose nice finish to the microscopic facet of the trilobite's eye, has for countless ages been preserved for us in this grand cabinet, unharmed by corroding elements, undisturbed by the

seas that have swept above them, and the forces that have broken and lifted the earth beneath them.

Then she traces the "footprints of the Creator" in the quarries of the Old Red Sandstone, and again wanders through the luxuriant forests of the carboniferous flora. Mounting one step higher, she splits from the Connecticut Valley the paths of gigantic birds, and forms unknown among living fauna. From the shores and waters of the Oölitic ocean she brings up reptilian life, in form more wonderful, and in armor and strength more terrific, than painter or poet ever dreamed of. And once more the earth almost seems to tremble beneath the tread of Mastodons, whose bones she brings up and places joint to joint in all their vast proportions. And last, above the tribes entombed in rock, she points to man, the crown of all—not only the last, but the most perfect being ever formed upon the earth—with all the faculties needed by a rational soul, showing, in his physical organization, that he is the last of the long series, and beyond him no progress is possible, according to the plan dimly sketched in the first vertebrate of the Silurian waters, and unfolded through all succeeding geologic ages.

As beneath the corner-stone of human structures are placed mementoes for coming generations, that they may know more perfectly the works of their fathers who reared the walls, so beneath the foundation stones of this earth have been deposited, by the Great Architect, the records of His works, for the study of him who was to be brought last upon the scene, the most perfect work of His hand.

In treating of the relations of Natural History to Religion, we are not disposed to ignore the fact that, from the progression of the *plan* of creation, from the simplest organic forms in the lowest rocks to the highest plants and animals of the present era, an argument has been drawn by some against the necessity of a personal God. Misinterpreting the evidence of progression of the plan as new species were introduced, they have applied the law of progression to single species, and thus are led to believe that the forms in the lower rocks have gradually changed, and in consequence of these changes that they present in their upward development all the

phases of life which Geology has revealed. Such a
strange reading of geologic text would yet require a
divine power to introduce the first germ of life.
But error seldom stops till it has reached the edge
of the precipice, and stepped over for its fatal fall.
There was a time, however, when Natural History
seemed ready to furnish the Atheist and Pantheist
with a magazine of missiles, against which the
strongest walls Religion has raised were doomed to
crumble. But they have been hurled back with a
force that has silenced the attacking batteries. The
most enticing, the most plausible, and the most dangerous was the theory of development, or transmu-
tation of species, making the sea, that fruitful field
of life, the birthplace of every organic being—not in
perfect form, but as a mere vital point where water
touched the land ; invoking for the original creation
no higher power than the electric current.

This theory of transformation was amusing, rather
than mischievous, as dreamed out by Maillet a hundred years ago. No serious harm could come from
fancying a shoal of frightened fishes floundering
among reeds till their fins were split into feathers,

and their noses lengthened and hardened into beaks, so that it should be more convenient for them to fly away and light on trees than to return to their native element. But in the hands of the able naturalists Oken and Lamark, it assumed a more scientific and more dangerous form. But their arguments, like boomerangs thrown by unskillful hands, have returned against themselves. The last blow was given by the stone-mason of Scotland. He thus describes the discovery that was to him the grand weapon of defence, and of carrying the war into the enemy's camp.

"The day was far spent when I reached Stromness; but as I had a fine, bright evening before me, longer, by some three or four degrees of north latitude, than the midsummer evenings of the south of Scotland, I set out, hammer in hand, to examine the junction of the granite and the Great Conglomerate, where it has been laid bare by the sea along the low promontory which forms the western boundary of the harbor. . . . . . I traced the formation upward, this evening, along the edges of the upturned strata, from where the Great Conglomerate leans against the granite, till where it merges into the ichthyolitic flagstones, and then pursued these from older and lower to newer and higher layers, desirous of ascertaining at what distance over the base of the system its most ancient organisms first appear, and what their character and kind. And, imbedded in a grayish-colored layer of hard flag, somewhat less than a hundred yards over the granite and about a hundred and sixty feet over the upper stratum of conglomerate, I found what I sought— a well-marked bone, in all probability the oldest vertebrate remains yet discovered in Orkney. . . . . The amateur geologists of Caithness and Orkney have learned to recognize it as the 'petrified nail.'"

To a looker-on, it would have seemed a thing of little importance, that evening-stroll of the lone geologist. But it was a memorable evening for science and religion. The blows of that little hammer are still sounding, and that "petrified nail" was more fatal to the development hypothesis than the tent-nail to the temple of Sisera.

It proved that the earliest fishes were among the highest organized; the order was reversed, the argument from development was broken. And his final language is fully vindicated: "They began to be, through the miracle of creation." Thus, then, from this apparent danger, has the human mind been quickened; and nature, summoned to testify against the existence of a personal God, has from the deep strata given such a response as proves His being to all the ablest geologists of the world.

In all I have thus far said of the bearing of Natural History upon Religion, I have not once referred to a written Revelation. But the time is past when it is considered out of place to refer with respect to the Bible, even in a temple of science. But how long would it maintain its power in this age, if it could be shown to contradict the teachings of science? It might exert a power in dark ages and among ignorant men, as the sacred books of other religions, the Koran and Vedas, have so long done. But the human mind must change before a book shall hold its sway against the teachings of Nature, as science now unfolds them. There are in science some things at least sure, and they must, and will have their weight. A book is written by men, it may therefore be simply a human production—it may be changed, portions lost, and portions added. But no human hand has lifted the hills, no scheming founder of religions has rolled back the strata of the earth, and placed there the fossils to mark the supernatural introduction of life. No company of his followers have set the forests, and filled the waters with their teeming tribes. He that believes in a God, believes in Him first as the creator, and will believe in no book as coming from Him, through the instrumentality of men, that contradicts this revelation of nature, which came direct from the hand of God, before the creation of man, and beyond the power of man to change in a single letter. If the Bible is the message of God, delivered by His embassadors, who were clothed with plenary power to do it, Nature is the autograph letter of the great Sovereign; and now, that men have learned to read it, they demand as condition of belief in the message written by men that it shall not contradict the letter, which they know to be genuine, stamped as it is with the great seal of almighty power which He has committed to the

keeping of no created being. The relations of that
portion of science called Natural History to the
Bible, may appear to some not of a marked and
direct character, but only incidental. But even
these incidental relations are of the highest value,
by throwing light upon obscure portions of the
inspired record, leading to more profound study and
more liberal views in its interpretation. That
religion which is worthy the name, is not secured by
simply proving the existence of a God, who might
have originated the universe, set it in motion once
for all, like a vast machine, and then withdrawn
himself forever from its government and special
care. It is in the revelation of character as con-
tinually guiding and caring for His creatures, that
the foundation is laid for rational religion, that
shall manifest itself in trust and action. This
character is certainly revealed in the creation, and
" may be understood by the things that are made."
And in all these works, with the manifestation of
intelligence is joined benevolence, so that like bina-
ry suns they move on together. The special care
manifested through all the geologic ages in fitting

each new species for the conditions of the globe
when it lived, makes it hard for us to believe that
this care has now ceased. We have at least the
proof that care has been exercised, not once merely,
but unnumbered times through the long lapse of
ages. If He cared for the fishes of the Silurian seas,
will He not care for us? The thousand miraculous
interpositions proved by the introduction of species,
and man himself, show that God introduces the
supernatural whenever the good of the universe
requires it. And now, when we see the careful
provision he has made for the wants of every created thing, it renders us more ready to see in the
adaptations of the Bible to the wants of man's
higher nature the mark of His hand.

We thus, from the study of Nature, remove all
antecedent probability against the Bible as a revelation, and against the miracles by which its divine
authority is supported. The supposed disagreement
of the two books has led to more careful study, not
only of the rocks, but of the Hebrew text, and its
influence on Biblical criticism is of the most marked
and happy kind. They may never be perfectly

reconciled; nor do we care for that. They move on in the same direction—both declaring a personal God, both declaring his miraculous interposition, both declaring his continued care. The intelligent theologian would be hard to find who does not understand that the Bible would lose its force if shown to conflict with science, and who does not know that the Natural History of the earth has destroyed infidel arguments which metaphysics could meet only by words.

Special adaptations and evidences of a divine interposition in distinct acts of creation, are sufficient perhaps for the intellect, but they are hardly of more importance than the adaptation of Nature to the emotions. "Blessed are the pure in heart, for they shall see God." In a lower sense this has its application, for the pure in heart are most ready to see the proof for the existence and attributes of God. If, then, nature is fitted to develop in man a true taste, giving him the types of the beautiful, it must purify and elevate the feelings, and prepare him for communion with the Author of Nature. Such can not fail to be its tendency.

We necessarily take man as our type of personality. Is it possible, then, we ask, to prove that
personality from any of his works? If this is
denied, then our argument from contrivance is
certainly in danger, because we have no acknowledged standard. But if it be granted that any work
of man, any of the grand material results brought
out by the combined wisdom and skill of the race,
proves personality, then we have a recognized
standard. Let that be taken, and I care not what
it may be, and it can not only be matched in every
particular in the works of Nature, but as far exceed-
ed in completeness as in grandeur and beauty. We
see one plan or set of plans commenced in the first
creation of animal and vegetable life, the grand
ideas in those plans preserved till the present
moment, for untold ages, not only through thousands of generations, but through thousands of new
creations. And yet that plan has been modified in
its details, to carry out a particular design in each
new species. This wisdom and skill are thus seen,
not only providing for the exigencies of the day,
not only for the things already created, but looking

forward through geologic ages of physical change, and providing for the wants of man, in his needs and desires, entirely unlike any thing before created.   It was for man alone that metals were poured into the primary rocks, even before life was introduced upon the globe; it was for his need the coal was garnered up, ages and ages before the earth was fitted for him.   We can hardly see a fold in the strata, or study a new form of matter, that does not seem to have reference to man as a physical or intellectual being.   But without going thus far, we can assert that he has been perfectly provided for; and *what short of the wisdom and skill of a personal being could provide for the wants of man, from whom alone we have our idea of personality?* While, then, we know the argument from mind must be satisfactory to the philosopher, we must also believe that the fitting up of a body for that mind, and a world for that body, are equally proofs of personality.   For none but a person can understand, so as to provide for the wants of personality. The chain then seems to be unbroken.   If the creation of the mind would prove personality, then

14

the body fitted to that mind—and if the body, then
every special adaptation by which that body is
adjusted to the forces of the natural world.

We believe this view is sustained by the Apostle.
"*For the invisible things of him from the creation
of the world are clearly seen, being understood by the
things that are made, even his eternal power and
Godhead, so that they are without excuse.*" Surely
it can not be contended that the Apostle supposed
any thing less than a personal God was manifested
by creation, when he declared that they were without excuse for not worshiping him as God. But on
the other hand, he denounces them for likening
God to *corruptible man, and to birds and four-
footed beasts and creeping things*, as though they
could discern in creation no higher marks of wisdom than instinct, or the imperfect works of sinful
man.

But let it not for a moment be thought that I
offer Nature as a substitute for the Bible, or the
love of God as the author of the beautiful as the
sum of that love demanded by Him as a righteous
moral Governor. Nature is a revelation, and if

rightly studied, so far from satisfying us, will teach us the need of another, higher and plainer. It begets the childlike spirit, teachable and pure—fitted to receive a full revelation, as the Bible claims to be, and to enter upon that life of faith which the Bible demands and the soul of man craves.

Nature and the Bible can each be studied alone; but as God is the author of both, we can never believe that the lowest can be neglected without loss, as we know the highest can not be without shipwreck of all the nobler objects for which man was created.

Their relation can not be better expressed than in the language of M'Cosh. "Science," says this able author, "has its foundations, and so has religion; let them unite their foundations, and the basis will be broader, and they will be two compartments of one great fabric reared to the glory of God. Let one be the outer and the other the inner court. In the one, let all look, and admire, and adore; and in the other, let those who have faith, kneel, and pray, and praise. Let the one be the sanctuary where human learning may present its richest incense as an offering to God ; and the other the holiest of all, separated from it by a vail now rent in twain, and in which, on a blood-sprinkled mercy-seat, we pour out the love of a reconciled heart, and hear the oracles of the living God."

The End.